# My Corydoras

by Frank Schäfer

Disclaimer of liability:
While every effort has been made to ensure that the information in this book is accurate, the author and publisher disclaim all responsibility for any errors. In purchasing this book the owner expressly accepts this disclaimer of liability.

All rights reserved. No part of this book may be reproduced, stored in any data retrieval system, photocopied or copied by any other method (electronic or otherwise), used for broadcasting or public performance, without the express permission of the publisher.

My Corydoras/Frank Schäfer. - Rodgau: A.C.S. (Aqualog minis); NE: Schäfer, F. (2003)
ISBN 3-936027-25-0
German edition: ISBN 3-936027-24-2
Dutch edition: ISBN 3-936027-26-9
Swedish edition: ISBN 3-936027-28-5
© 2003 by Verlag A.C.S. GmbH (AQUALOG), Liebigstraße 1, D-63110 Rodgau /Germany
phone: +49 (0) 6106 690 140
fax: +49 (0) 6106 644 692
e-mail: info@aqualog.de
http://www. aqualog.de

Text, research, editing, photo selection, layout, and titling: Frank Schäfer
English translation: Mary Bailey
Business Manager: Ulrich Glaser sen.
Printing, typesetting, processing:
Lithos: Verlag A.C.S.
Printing: Westermann Druck Zwickau
Printed on Magnostar gloss, 100% chlorine-free, environmentally friendly paper.
PRINTED IN GERMANY

All illustrations are from the Aqualog photo archives, except page 6 (artist unknown, from Zernecke, "Leitfaden für Aquarien- und Terrarienfreunde", 1899)
All enquiries regarding photos should be directed to Bildarchiv Hippocampus, www.hippocampus-bildarchiv.de

Photographers: Dieter Bork, Ulrich Glaser sen., Steffen Hellner, Burkard Migge, Schuzo Nakano, Jens Pinhard, Frank Schäfer, Erwin Schraml, Frank Teigler, and Tomizana

## Contents:

| | |
|---|---|
| Masters of defense | 4 |
| Pioneers of the aquarium hobby | 5 |
| The scientific species name | 7 |
| What is a mailed catfish? | 8 |
| The parts of a fish | 9 |
| Gut-breathers | 10 |
| The mailed catfish aquarium | 11 |
| How large an aquarium? | 13 |
| Biological equilibrium in the aquarium | 14 |
| How a filter works | 15 |
| Plants in the aquarium | 16 |
| Plants in the mailed catfish aquarium | 17 |
| Planting plants | 20 |
| Water for the mailed catfish aquarium | 21 |
| Chemistry – how water works? | 22 |
| Feeding mailed catfishes | 23 |
| Disease in mailed catfishes | 24 |
| Copy-cats?! | 29 |
| Species trio 1 | 30 |
| Species trio 2 | 34 |
| Shoaling fishes? | 38 |
| More mimicry | 39 |
| The Paraguay connection | 40 |
| Floating eggs – the *C. elegans* group | 42 |
| Mini-corys | 44 |
| Hatching *Artemia* | 45 |
| Cool sideburns | 46 |
| Gleaming metal | 48 |
| Round heads and high backs | 50 |
| Black and white pandas | 52 |
| Evolving sand-dwellers | 54 |
| Oddball corys (1) | 56 |
| Oddball corys (2) | 57 |
| Individualists | 58 |
| Specks of gold | 60 |
| Pet leopards | 62 |
| Born mimics | 64 |
| Poisonous spines | 66 |
| All good things | 67 |
| Breeding and rearing | 68 |

## The blue pages

On the blue pages you will find general information relevant to the aquarium hobby. If you already have fish-keeping experience you can skip them. But we have chosen this method because we have no idea what you already know, and constant references to other literature are not much help.

# Masters of defense

Life is not easy for a small fish. Death and disaster lie in wait for it everywhere. Enemies that would like to eat it are legion: other, larger fishes, aquatic reptiles (turtles, snakes, lizards, crocodiles), fish-eating birds, even large arthropods, such as certain spiders, but also shrimps, aquatic beetles, etc, etc.

It is thus hardly surprising, that small fish species need to have plenty of tricks up their sleeves in order to be able to survive. Some of them have learnt to fly, for example, the hatchetfishes, members of the characin family, which escape many predators by boldly leaping out of the water. Others – and very many small fishes do this – seek safety in numbers, forming large shoals with conspecifics and similarly-colored species. Statistically speaking, they thus have a better chance of it being another member of the shoal that succumbs when a predator attacks. At the same time it is more difficult for the predator to single out any individual fish in the shoal. Yet other small fishes seek out the protection of the night. A number of auchenipterid catfishes squeeze themselves into narrow cracks in wood and crannies among rocks during the day, coming out only at night to go about their business. In the aquarium, this means that these very pretty little catfishes are only to be seen if you come home late at night. *Corydoras* and other mailed catfishes have adopted a number of these tactics simultaneously. They swim in shoals, often mixed with other, similar, species. Many recent studies indicate that, in addition, they employ ingenious migratory techniques, for example spending the day in deep water and the night in the shallows. However, none of the estimated approximately 300 mailed catfish species has as yet learnt to fly! Instead, however, they have evolved a defense whose perfection is unmatched in any other fish species: a bony armor surrounding the body, which, together with the sharp first spines of the dorsal

Hatchetfishes have learnt to fly.

and pectoral fins, make sure that they stick in the throat of almost any creature that tries to swallow them. These defensive tactics have made mailed catfishes exceptionally successful, and there is hardly a body of water in South America that is not inhabited by one species or another.

Auchenipterid catfishes are often very attractively colored, but in the aquarium these highly nocturnal creatures rarely show themselves.

# Pioneers of the aquarium-hobby

Mailed catfishes, or more accurately, the peppered cory, Corydoras paleatus, were the third exotic fish species to be imported to Europe especially for aquarium maintenance. They were preceded only by the goldfish and the paradise fish (*Macropodus opercularis*). In 1876 Captain Rousseau brought the first specimens to Paris, where they were handed over to the legendary fish breeder Carbonnier.

At that time it was the height of fashion to bring exotic animals and plant species to Europe, and, if possible, naturalize them. This was regarded as enriching the local flora and fauna. During this era nature was viewed with a remarkable mixture of romanticism and the pragmatism of the industrial revolution. Hardly anything was known about ecology, the study of the delicately intertwined relationships between all life forms and with the natural, unpeopled, environment that surrounded them. It was known, however, that alienating people from this natural world had harmful effects on their physical and mental health. The aquarium hobby, being nature-oriented, became popular at this time, not only among the educated citizenry but also with the "man in the street", and clubs were formed in industrialized nations worldwide in order to propagate this new form of animal maintenance.

The peppered catfish, one of a very small number of species with a widely-accepted common name, was discovered by no less a person than the famous Charles Darwin, during the legendary voyage of exploration of the Beagle, during which Darwin developed his theory of the evolution of species via mutation and selection. The fishes collected during this expedition were studied by Darwin's colleague Jenyns, who described the peppered catfish as *Callichthys paleatus*. *Callichthys* derives from the Latin word callum meaning hard skin and the Greek word ichthys meaning fish, referring to the external armor of these creatures. However, the genus *Callichthys* contains a single species (more of this later on); the species *paleatus* was subsequently transferred to the genus *Corydoras*, erected by the French ichthyologist Baron de la Cépède in 1803. A historical footnote: La Cépède's book on the natural history of fishes was published at

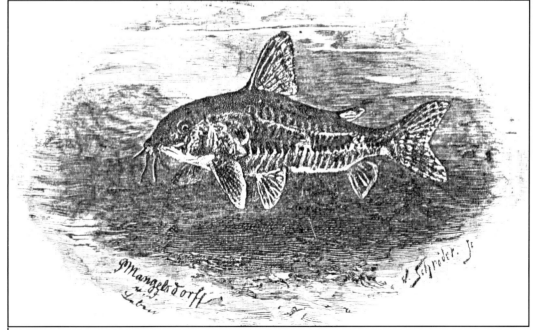

The peppered cory, *Corydoras paleatus*, from an 1899 aquarium book.

the time of the French revolution. "La Cépède" means "the citizen", but I have no idea whether the Baron saved his skin by changing his name or whether he was genuinely a supporter of the revolution. It is a fact, however, that he himself was not quite sure how he should now write it. It is written in a huge variety of ways in the various volumes of his book, sometimes also as the single word Lacépède, and to the present day there is considerable confusion as to how to cite this very important work. I am quite sure that not all of my colleagues will agree with my preferred usage. But that is by the by. Back to the peppered catfish. The word paleatus derives from the Latin palea meaning chaff, probably referring to the pattern of scattered spots in this species.

As early as 1878 CARBONNIER succeeded in breeding this species. In 1893 the first peppered corys were brought to Germany from France by an enterprising Berliner, NITSCHE.

It was a stroke of luck for the aquarium hobby that it was this relatively southerly-distributed (in Argentina and Paraguay plus the Pantanal in Brazil) species that reached us first, as these fishes will tolerate relatively low temperatures – a prerequisite for the long-term establishment of any aquarium fish given the sometimes "adventurous" heating methods of our forefathers! In addition, *Corydoras paleatus* is one of the easy-to-breed corys. Even experienced breeders are still gnashing their teeth over a number of other species.

A second southerly species, namely *Corydoras nattereri* (named after NATTERER, a famous explorer of South America), followed at the turn of the century (19th/20th). Since then a succession of mailed catfish species have been imported, their identification sometimes presenting huge difficulties.

*Corydoras nattereri.* This species from southern Brazil found its way into the aquarium early on.

# The scientific species name

The scientific classification of the animal and plant kingdoms requires a unified system of nomenclature. Obviously animals and plants also have common names in their natural distribution areas, but these names are not governed by any rules, and so widespread species also have numerous popular names. Take the plant shown here.

It grows all over Europe, and I first encountered it as rabbit food! The English call it dandelion, a corruption of the French dent de lion, and in Germany it is Löwenzahn, in both cases meaning "lion's tooth", referring to the jagged-edged leaves. But the French also call it pis-en-lit, referring to its diuretic properties, and the Germans have a host of other names: Milchbusch (milk bush), from the milky sap that oozes from broken stems; Pusteblume (blow-flower), from the fluffy ripe seed-heads which it is such fun to blow away; and Butterblume (butter flower), a collective name for yellow-flowered plants of summer meadows.

All these names for one species, and I am sure there must be more.

It was obvious that the scientific classification of the animal and plant species of the world required a system that would permit understanding regardless of geographical boundaries. Such a system was devised by the Swedish biologist CARL VON LINNÉ (who, following the fashion of his time, preferred the Latinized version of his name, CAROLUS LINNAEUS). His basic idea was as simple as it was ingenious. First, he decreed, to be internationally valid names must be given in a dead language, as only names with no national connotations would gain international acceptance.

In addition, these names must consist of two parts. This system had already been tried and tested in the naming of humans, where each person has a personal and a family name. Linné therefore decided that closely related species should be grouped together in genera. Thus each species has a genus name followed by a species name. The genus name begins with an upper case (capital) letter, the species name with a lower case one. In addition, the name of the describer follows the species name, followed by a comma and the year of the original description. This makes it possible to locate the work in which the species was described. Thus the plant shown here is scientifically known as *Taraxacum officinale* LINNÉ, 1758.

# What is a mailed catfish?

Before we go any further, I think it is time to take a quick look at the rich and varied world of the mailed catfishes. It is well known that biologists catalogue animal and plant species using a system designed to clarify the various relationships between them. Fishes, just like us humans, belong to the chordates (phylum Chordata), but thereafter we go our separate ways. Humans belong to the class of the mammals, mailed catfishes to the bony fishes. Humans belong to the order of the primates, mailed catfishes to the Ostariophysi. The term requires a little further explanation. The Ostariophysi have a common characteristic, the so-called Weberian apparatus, a complex structure composed of various small bones (once upon a time the first four vertebrae), connected to the swim-bladder at one end and the labyrinth (part of the ear, and the organ of balance) at the other. This permits the smallest changes in pressure to be detected, and means that, quite simply, these fishes have very good hearing. The whole arrangement is so complex that it is thought improbable that it could have evolved independently on more than one occasion. But enough of that.

Within the Ostariophysi, the mailed catfishes belong to the sub-order Siluriformes where they form the family *Callichthyidae* (mailed catfishes!). This in turn subdivided into the sub-families *Callichthyinae* (whose members are sausage-shaped) and the *Corydoradinae*.

I realize this may all seem very dry and dull, but later on we have more than 150 scientifically described species, and almost as many scientifically undescribed species, to deal with. And so it is important to know what we are talking about, for in the final analysis every species of animal is unique in the truest sense of the word. That is to say, they all differ from one another in some way – otherwise we wouldn't know that they were different species!

The Ostariophysi do not all look the same. Their only common feature is the Weberian apparatus. From top to bottom: a knife eel (*Apteronotus albifrons*), a characin (*Paracheirodon innesi*), a cyprinid (*Carassius auratus*), and a catfish (*Platydoras costatus*).

# Gut-breathers

The heading of this chapter is in no way a delicately-phrased suggestion that mailed catfishes suffer badly from flatulence! But they do have a special adaptation allowing them to cope with oxygen shortages that sometimes occur in their natural habitat.

At such times they rise to the surface, gulp in air, and allow it to pass into the digestive tract. The mucus lining of the gut is well-endowed with blood vessels, allowing them to extract oxygen from the atmospheric air that they can then utilize. However, this does not mean that in an aquarium context these fishes should be kept in the equivalent of a cess-pit. Some especially hardy species, for example the bronze catfish (*Corydoras aeneus*), may well survive such treatment, but it certainly won't do them any good.

Although, as far as is known, all mailed catfishes have this ability to breathe via the gut, in nature they do so only rarely, when it is necessary. Only during periods of extreme drought, when fishes are crammed together in residual pools and must fight just to survive, is this auxiliary breathing method brought into play, giving mailed catfishes a decisive advantage over fishes lacking this ability. Even so, observant aquarists will find in these catfishes a living barometer. Their behavior changes as atmospheric pressure falls, and the frequency of their air gulping increases. In this way they forecast periods of bad weather.

Other bottom-dwelling and gut-breathing Ostariophysi, the *Misgurnus* loaches, have been given the popular name "weather fish" on account of this characteristic, and were once, like the tree frog, kept in primitive earthenware containers to prophesy the weather.

Bronze catfishes (*Corydoras aeneus*) are among the hardiest of mailed catfishes and suitable for beginners too. Keen observers can use them to forecast the weather!

# The mailed catfish aquarium

*Corydoras* are most often kept in so-called community tanks. One should take care that the correct substrate material is used!

Not all mailed catfishes are the same, a point that will be made repeatedly in this book. However, the majority of them have a number of maintenance requirements in common, that must be borne in mind when setting up the aquarium. For example, apart from two species, all mailed catfishes are bottom-dwelling fishes. So what does this signify in practice?

First of all, that the depth of the aquarium plays a very subordinate role for these fishes. The area of open bottom surface is far more important. The ideal mailed catfish aquarium should therefore not be crammed full of decorative objects such as bogwood, stones, and even plants. Open sand would most closely correspond to the natural habitat, but that would, of course, make for a rather boring aquarium. Nevertheless, if you want to observe many of the behavioral characteristics of these fishes, arrange the aquarium for maximum visibility.

The commonest error made during setting-up is incorrect choice of substrate material. Mailed catfishes are adapted for life on a sandy bottom. Coarse, in the worst case even sharp-edged, gravel (for example, lava gravel) will result in the fishes completely abrading away their delicate barbels, which are their most important sensory organ. They use these barbels to taste, to feel out their surroundings, and even to communicate among themselves, for example when mating.

In addition, mailed catfishes feed on fine food particles, spending the whole day foraging. This preferred type of food will rapidly disappear into the gravel where it will be inaccessible to many species. By contrast, the particles will remain lying on sand, and accessible to the fishes.

# The mailed catfish aquarium

But not all sand is the same, either. The quartz sand sold in the building trade is just as unsuitable for mailed catfishes as is coarse gravel. This sand is often ground up in gigantic mills, and the result is that the individual grains are pointed and sharp-edged.

This sand acts like emery paper, perpetually injuring the catfishes. Bacteria may then enter via the tiny wounds, and the result is sick and dying catfishes.

Anyone with no reliable dealer who stocks the right kind of sand should obtain some so-called "children's play sand". Details of possible sources can often be obtained from the local government department responsible for nursery schools. This play sand is obtained from sand pits in deposits laid down by rivers and/or glaciers long ago. It is round-grained and fulfils all the requirements expected of good aquarium sand.

The majority of plants grow only poorly in pure sand, but only a few aquarists will be prepared to forego a healthy growth of plants. Plant species that will thrive even in pure sand are covered in the next chapter. But there are also tricks that can be used to get round this problem. The simplest method is to build up the substrate in layers. The bottom 6 cm can be perfectly ordinary aquarium gravel, enriched with a long-acting fertilizer if desired. This layer is then topped with about 3 cm of fine sand, washed perfectly clean. The disadvantage of this tactic is that in the course of time the fine sand will work its way down and mingle with the coarser lower layer. Hence this type of mailed catfish aquarium has a "life" of only about two years before it needs to be set up again.

A second method is to grow plants that need plenty of nutrients in pots or trays containing the type of substrate they need, bedding these containers in the sand. The rims of the containers should be covered with stones so that they are not constantly exposed by any moving around of the sand. In this way the aquarium can be kept running undisturbed for a very long time.

The third, and very appropriate, method is to stick two or three strips of glass, about 8 cm high, to the bottom of the aquarium using silicone sealant. The zones behind these can be filled with the usual, enriched, aquarium gravel and planted exactly as normal. The foreground is then filled with sand and remains unplanted, thus creating an excellent mailed catfish habitat. Because the majority of these catfishes originate from flowing waters, they greatly appreciate a current from the filter. With this in mind, the filtration system should be such as to turn over the net volume of the aquarium about twice per hour. At the same time it is important that the filter inlet is not positioned too low in the aquarium, as otherwise a lot of sand may be sucked in, and too much of the food will disappear into the filter. If this results in a certain amount of mulm lying around the aquarium, that will do no harm. Mulm is a biologically highly active mass of decaying plant material, excreta, etc, and is an important nutritional element for mailed catfishes. Which is not to say that they can survive on a diet of rubbish!

It is very difficult to cite a minimum tank size for mailed catfishes, as the individual species differ markedly in their space requirements. The majority of species normally available in the trade will do well in a tank 60+ cm long and 30 cm wide, but the large-growing members of the genus Brochis, as well as the Pantanal cory (*Corydoras pantanalensis*, "C6"), need tanks with a bottom area of at least 120 x 50 cm. By contrast, a bottom area of 40 x 20 cm is perfectly adequate for dwarf species such as the pygmy cory (*Corydoras pygmaeus*) and the tiny false cory (*Aspidoras pauciradiatus*).

# How large an aquarium

In aquarium circles there is a saying, "An aquarium can never be too large." How so? Well, the main reason is that the body of water in the aquarium is more stable, chemically speaking, the larger its volume. In other words, a large aquarium is significantly less work than a small one.

Beginners and non-aquarists often think that fishes feel like prisoners in a small aquarium. This is not the case. Compared with the wild, even a large aquarium is no more than a tiny puddle, but fishes have no more concept of freedom than do other animals - such an abstraction is of no biological relevance to them. Only mankind has an inbred desire for freedom, and even so the concept has no single definition. Just ask 10 people of your acquaintance, what they understand by "freedom". In all probability you will get 10 different answers. In actuality, Man's quest for freedom is the recipe for his evolutionary success. It is nothing more than an innate feeling of discontent with the individual's personal circumstances, in consequence of which, depending on the degree of dissatisfaction, the person concerned seeks for an opportunity to alter his or her situation. By virtue of his inventive genius Man can adapt his environment to his needs and hence survive literally anywhere. In short, the human quest for freedom is a natural species-specific survival factor.

By contrast animals, including all fishes, are incapable of adapting their environment to their needs. Instead they are dependent, for better or worse, on their ability to adapt to their present environmental conditions. A blenny that decided to abandon its troglodytic existence for the lifestyle of a herring would survive only a few hours. Thus animals have no freedom. And thus the question of how large an aquarium has nothing to do with the amount of space available to a fish in nature. The aquarist should instead ask, "Would the fish species that I want to keep colonize my aquarium if the habitat it provides occurred in the wild?"

The significant differences between an optimal aquarium and the wild are: there are no enemies; there is an unlimited food supply; there is no competition; there are no natural catastrophes (drought, flood, etc) - the aquarist ensures all these things.

Accordingly the tank size necessary is a function of the expected eventual size and the behavior of the fishes. For inactive predators, that spend the entire day lying motionless in wait for prey, the tank length should be about three times, the tank width about twice, the body length of the fish. For active shoaling fishes the rule of thumb is a tank length at least 10 times body length, and tank width five times. Finally, the number of fishes must be taken into account. And here the old aquarists rule remains valid - at least two liters of water per cm of fish length.

If you are thinking of setting up an aquarium, please always bear its maintenance requirements in mind. Every aquarium requires a weekly or fortnightly partial water change of 10-25% of its volume. This removes pollutants, the accumulated waste products from metabolic processes, and also replaces depleted trace elements. For a 1000 liter aquarium that means 200-500 liters of water to be shifted (100-250 liters out, and the same back in). As a new recruit to the hobby you will do best to start with an aquarium of 150-300 liters capacity. A tank of this size will provide a chemically very stable volume of water and is a good size for almost all the aquarium fishes normally available in the trade.

## Biological equilibrium in the aquarium

It is impossible to achieve a true biological equilibrium in the aquarium - that must be understood right from the start. The amount of extraneous nutrients (in the form of fish food) is simply too great. One can - and should - nevertheless endeavor to create a stable aquarium environment which must then be supported by partial water changes and filter cleaning.

The basic prerequisite for such a stable aquarium environment is the use of water of consistent initial quality. That is to say, that the water used for changes should be identical in hardness and pH with that already in the aquarium. So think this over carefully before deciding on "home-brewed" water instead of your mains water. Because you will have to prepare it yourself, week-in, week-out!

The aquarium hobby is essentially about culturing bacteria. Without these invisible helpers it is impossible to run an aquarium. On the one hand there are the nitrite-forming bacteria. Fishes constantly excrete highly toxic ammonia (from the breakdown of protein) from their gills, and this first group of essential helpful bacteria convert this ammonia into nitrite (still highly toxic, and normally lethal to fishes in concentrations of 1 mg/liter). The bacteria need oxygen to convert ammonia to nitrite, and hence are termed aerobic bacteria. The second group of bacteria that make it possible for fish to live in aquaria are also aerobic, and convert the still highly toxic nitrite into relatively harmless nitrate. When maintaining an aquarium the aim should be a nitrate level of about 30 mg/liter; the value should never be appreciably higher, but lower is OK.

It is always the same genera of bacteria that perform this important nitrification, but every aquarium will have its own "micro-climate" depending on its basic water parameters, i.e. hardness and pH. The bacteria are appreciably more sensitive than fishes to fluctuations in water parameters, hence it is immensely important always to use "matching" water for water changes.

As well as the aerobic nitrifying bacteria there are innumerable other micro-organisms, i.e. bacteria, fungi, etc, that colonize aquaria. The higher the nutrient loading of the aquarium, the higher the number of these organisms in the water. Additional factors that increase the micro-organism population include the fish population density and the amount of convertible organic material in the aquarium, i.e. the so-called mulm. Mulm consists of fish excreta, dead vegetation, uneaten food, etc, and it makes no difference whether it is lying around the tank or out of sight in the filter! The organisms that process mulm are intrinsically harmless, but if their population increases to excess then these normally harmless organisms represent a danger to the fishes. The immune systems of many fishes kept in aquaria are naturally only weakly developed, as the micro-organism population in many tropical waters is extremely low because they are very nutrient-poor. Hence it is obvious that the micro-organism population should be kept as low as possible by siphoning off mulm during water changes, filter maintenance, and sensible limitation of the fish population.

Should it be necessary, for whatever reason, to populate the aquarium densely with fishes, then a UV sterilizer, installed in the filter return, is one way of effectively reducing the number of micro-organisms in the water. But – and this is where skill comes in – the aquarist should always seek to create a degree of biological equilibrium through knowledge and thought, and use special equipment only where unavoidable.

# How a filter works

There are many different types of filter, all with advantages and disadvantages. Basically, every filter has a mechanical and a biological section. The former serves to remove particles – that cloud the water or are regarded as dirt - from the aquarium. For this purpose the aquarium water is drawn through a suitable filter medium and the cleansed water is then pumped back into the aquarium. This mechanical cleansing is usually achieved using filter floss, sponge, or the like. You should get into the habit of cleaning this "dirt filter" weekly during the partial water change. Ideally the filter medium should be rinsed in a bucket of newly siphoned-off the aquarium water, as this will avoid harm to the useful aerobic bacteria also contained in the medium.

The biological section is usually divided into various zones. The best-known is the aerobic zone, which endeavors to produce the largest possible population of the aerobic bacteria that convert ammonia to nitrate via nitrite, by providing a substrate with the largest possible surface area for colonization. This process is strictly oxygen-dependent. Typical substrates include ceramic tubes, "bioballs", various artificial materials, porous clay balls, and even basalt chips. The most extreme form of the aerobic filter is the so-called trickle filter, in which the water cascades over thin layers of filter media and is thus constantly supplied with abundant oxygen. This works exceptionally well, although only miserable plant growth is possible when the water is processed this way, and in addition the high oxygen level encourages rampant algal growth. For this reason such filters are best used only for heavily populated aquaria with alkaline water, where the danger of ammonia toxicity is greatest.

More and more frequently the aerobic filter indispensable in any aquarium is nowadays complemented by an anaerobic filter, inhabited by anaerobic bacteria to which oxygen is toxic.

These filters have two great advantages. Firstly they can be used to cultivate bacteria that break down the relatively harmless nitrate into gaseous nitrogen and oxygen. Both gases can then escape from the water. If the filter functions properly then it can thus be used to keep the level of nitrate in the aquarium very low. Secondly, in this type of filter the highly important plant nutrients that are oxidized by aerobic filters (and thus rendered useless to plants) undergo reduction, i.e. the oxygen is removed again. For this reason many aquarists run a slow-flow anaerobic filter in bypass, i.e. connected to the outlet of the aerobic filter.

There are various media for anaerobic filters. Special artificial media are used for the nitrate reduction filter, sold already impregnated with the required bacterial culture. As a rule a small external filter with a low throughput is used for the plant-friendly bypass filter, and filled with, for example, fine sand or special filter media such as sintered glass, etc. Your dealer will be pleased to advise on this.

The filter can also contain materials with a special purpose, for example filter carbon. This so-called activated carbon is very effective for removing some medication residues from the water, as well as yellowing and other types of discoloration. Filter carbon should be used only for a specific purpose, not on a permanent basis. In addition a bag of peat can be placed in the filter to acidify the water. There are also special ion-exchange resins, which, when necessary, can be used to bind up nitrate or phosphate, lowering levels of these pollutants quickly and effectively. Special filters filled with diatomaceous earth can be used to produce sparkling water and even reduce the number of microorganisms in the water.

# Plants in the aquarium

Plants perform a number of functions in the aquarium. They remove pollutants, which serve them as food for growth. They provide hiding places for harassed adults and newborn fry. They produce oxygen and reduce carbon dioxide levels during the day. And finally, a planted aquarium is much more attractive than a bare tank.

It is best to choose large, broad-leaved for the mailed catfish aquarium. Choice of the correct substrate is important for a planted aquarium - it must be at least 7 cm, better 10 cm, deep.
At the bottom you can use a 1-2 cm deep layer of "compost" material containing a long-term mineral fertilizer in the form of special clay or similar. Horticultural fertilizers and potting composts are totally unsuitable as these have a high content of organic nutrients that will pollute the aquarium and ruin the water quality.
On top of the compost there should be a 5-7 cm layer of fine aquarium gravel, which should, as usual, be thoroughly washed before use.

Finally there is the top layer, which should again be painstakingly washed clean. Normal aquarium gravel will be fine. However, as mentioned earlier, there must always be areas of fine sand that will be the actual habitat of the mailed catfishes in the aquarium. The sand too must be meticulously cleaned.

Next carefully add just enough water to thoroughly moisten the substrate. This will enable you to plant the aquarium without a load of "dirt" floating up from the bottom layers. Once all the plants are in position, you can - carefully! - fill the aquarium with pre-warmed (to 18-22 °C) water.
Finally set the filtration and heating going. Now the aquarium should remain unoccupied for at least two weeks - this is necessary to allow the aquarium to develop the population of micro-organisms needed to turn a sterile tank of water into a functional aquarium biotope. During this period the plants will start to grow and so the lighting must be turned on right away.

This is what an aquarium should look like if it is to satisfy the corys' need for open spaces with a fine sand substrate.

# Plants in the mailed catfish aquarium

Mailed catfishes do not eat plants and as a rule do not nibble them at all (the only exceptions are the large *Megalechis* species (formely *Hoplosternum*) during nest construction). Thus we can, in principle, grow any plant in the mailed catfish aquarium provided its temperature requirements match those of the fishes. However, light-loving species are best avoided as mailed catfishes do not like excessively bright lighting.

There are, however, a number of plant species that can be considered poorly suited to the mailed catfish aquarium. For example, milfoil (*Myriophyllum*), which firstly does not like currents, secondly needs a lot of light, and thirdly tends to collect fine particles stirred up by the catfishes on its leaflets, causing it to die back. The same applies to *Cabomba* and *Limnophila*.

Ideal plants for the mailed catfish aquarium include the African *Anubias* species and Java fern (*Microsorium*) from Asia. These plants are quite happy with relatively little light and can be grown on stones or bogwood. They can thus thrive in an aquarium with a substrate entirely of sand. Also well suited are the *Najas* species, and hornwort (*Ceratophyllum*). Both these stemmed plants do need rather more light than the previous two, but they derive their nutriment directly from the water and form few or no roots. They can thus be pushed straight into pure sand. Even if they are "uprooted" by the catfishes, they will continue to grow, floating free in the water. Unfortunately floating plants do poorly in a tank with plenty of water movement. But during the simulation of a drought period, which is necessary for the breeding of many species (more of this later), the floating form of Indian fern (*Ceratopteris cornuta*) is an excellent plant for the mailed catfish aquarium.

We will now take a brief look at these particularly suitable plants:

**Dwarf Anubias (*Anubias barteri* var. *nana*)**

This attractive plant is one of the easiest aquarium plants of all, and can be unreservedly recommended for mailed catfish aquaria. It does well attached to decor items (see "Planting plants") and will grow in acid or alkaline water, and even in dim light.

Unfortunately this plant grows rather slowly, so that it is usually rather expensive to buy. Mailed catfishes are very fond of spawning on the tough, leathery leaves of this plant, so it is well suited to the breeding aquarium. The different varieties of this species differ mainly in their height: *A. barteri* var. *nana* produces leaves only 3-5 cm in length.

# Plants in the mailed catfish aquarium

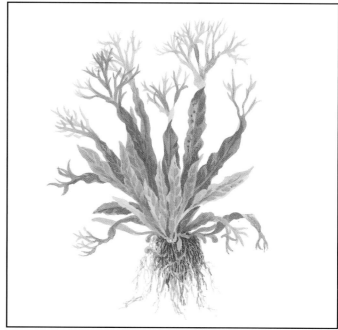

**Java fern (*Microsorium pteropus*)**

Generally speaking the same applies for Java fern as for Anubias: this plant too will grow almost everywhere and anywhere and does well attached to rocks, etc. This fern is propagated via small plantlets that form along the edges of older leaves.
There are several very attractive cultivars of Java fern, e.g. the "Windelov" and "Tropica" forms. Unlike the Anubias, which develops very sturdy clinging roots, the roots of Java fern are fine and form a dense, black-brown mat. Some mailed catfishes, especially dwarf species, like to spawn among these roots.

## Plants in the mailed catfish aquarium

*Najas guadelupensis*

Hornwort *(Ceratophyllum demersum)*

This plant is very popular with fish breeders as it will grow in almost any water and can even be cultivated free-floating, without substrate. Its light green leaves form an attractive contrast with the previous two species.
Unfortunately the stems are very fragile, like glass, for which reason many aquatic plant nurseries have little time for this plant - too difficult to transport. However your dealer may nevertheless be able to obtain a few plants for you. If not, try your local aquarium club.

Hornwort is a stemmed plant that never forms roots, hence it can be grown loose in the water. It will, however, also tolerate being pushed into the substrate. Hornwort is very useful in the breeding aquarium and provides outstanding shelter for fry.

Hornwort and Najas are very fast-growing and hence remove large amounts of wastes from the water. Hence no mailed catfish aquarium should be without at least one of these plants to help maintain the biological equilibrium.

# Planting plants

Basically, three types of plants are cultivated in the aquarium: floating plants, stemmed, and rosette-forming plants.

The floating plants are most easily "planted". They are simply placed on the water's surface, and all that is necessary is to make sure the roots are pointing downwards - and even that is superfluous with the rootless types. All floating plants "dislike" filter currents, and it is often pointless to try and grow them in heavily filtered aquaria.

Stemmed plants form only a moderately extensive root structure. They are propagated via cuttings taken from any stems that are long enough. As a rule cuttings should be about 10 cm long. When planting stemmed plants the following points should be noted: never plant them in bundles, but insert each stem separately. The lower leaves should be removed, as if buried in the substrate they will rot and possibly infect the stem with decay. When you buy stemmed plants from an aquarium store, they will usually be clipped together with lead or planted in a small pot - the lead or pot should be removed before planting. Finally, remove any roots already present. If the lower stem looks transparent then it has been squashed - cut off the affected part with a sharp knife before planting the remainder.

Finally, the third group of plants comprises the rosette-forming types. These plants form an extensive root structure. They are propagated via runners or offsets. With these plants too, any lead strip or pot must be removed before planting, and the roots should then be gently teased out and shortened to about 3 cm long using a sharp knife. When planting, it is essential that the roots all point downwards into the planting hole - if they get bent upwards during planting then the plant will not grow well. Rosette-forming plants possess a "crown" or woody rootstock (rhizome) from which the foliage grows, and it is important to ensure that this growing point is not buried in the substrate but extends a few millimeters above it.

A number of rosette-forming plants of the genus *Anubias*, as well as ferns of the genera *Microsorium* and *Bolbitis*, do not grow well if planted in the substrate. These plants are best tied to wood or porous stones using dark cotton, and will attach themselves firmly in time. These plants can often be purchased growing on rocks or wood.

Many rosette-forming plants are marsh plants by nature, and grow submerged only part of the time in their natural habitat. To this group belong many species of the genera *Cryptocoryne* and *Echinodorus*, for example. When first planting an aquarium, these plants should generally make up only about a third of the species used. This is because these plants grow only relatively slowly. During the initial phase of a new aquarium the biological conditions are such that many undesirable algae find an optimal environment for growth.

Because aquatic plants and algae compete with one another for resources, logically fast-growing plants will be more successful than slow-growing ones. Your dealer will, of course, be pleased to advise which plants will be suited to your aquarium. But if you want an extensive discussion with your dealer it is best not to visit him during his busy period - this also applies when the object is to design a planting scheme.

# Water for the mailed catfish aquarium

To date mailed catfishes have proved delightfully adaptable, at least where general maintenance rather than breeding is concerned. Although in nature each species is adapted to its own special environment, even those that live in exceptionally soft water do well long-term in medium-hard and slightly alkaline conditions.

Mailed catfishes are residents of tropical and subtropical South America. The family is thus very widely distributed. Two of the main types of water found in South America have been colonized: the sediment-rich whitewaters, which are so murky that visibility is often only a few centimeters; and clearwaters, whose clarity is reminiscent of a number of mountain lakes. On the other hand, they are not found in the dark, cola-colored blackwaters, which contain little or no suspended material but are so dark in color that underwater visibility is severely limited. There are all sorts of intermediates between these three main water types, but for this classification system has proved perfectly good for practical purposes. From a chemical viewpoint the whitewaters most closely resemble what comes out of the tap in many parts of central Europe. Depending on the time of year there may be measurable hardness and the pH value may rise noticeably above neutral (up to 7.5). Fishes from whitewater biotopes are thus often relatively easy to breed in the aquarium, as they often come into breeding condition even in the general community.

Black- and clearwaters, by contrast, are characterized by a lack of hardness and a very acid pH, often as low as 4.0-4.5.

Species from black- and clearwaters will tolerate high micro-organism counts in the water only poorly, rapidly reacting to this situation with shimmying body movements, clamped fins, and loss of appetite. In such cases you must always immediately perform a large water change and eliminate the root cause. Too much food? Clogged filter? Clogged airstone? Always check the nitrite level. A clear sign of nitrite toxicity is accelerated breathing and when the catfishes keep turning on one side and scraping their gillcovers on the bottom.

The special requirements of individual species and groups of species will be covered later, but it must be made clear that no "patent recipe" can be provided where there are so many species and so many natural habitats.

However, two special scenarios must be mentioned. If you have a domestic water-softening unit in your house, the water from will be unsuitable for guppy maintenance. In such cases you must take the water for your aquarium from the mains, before it passes through the unit. If you have new copper pipework then the water from them will initially be suspect, as they need to develop an internal layer of oxide before the water is usable. Copper is highly poisonous to fishes.

Corys often have gleaming markings that may serve to keep the shoal together. This is *Corydoras adolfoi*.

# Chemistry – how water works

Even if you have previously regarded chemistry as not your particular cup of tea, a few basic elements of chemical knowledge will not come amiss in the aquarist.

First of all there is water hardness. Most people will already have heard of this, as water hardness is responsible for the "chalking up" of kettles, hot water pipes, etc. The concept of water hardness originates from the washing powder industry and was originally used to quantify the amount of soap powder needed to create an effective lather for washing. Only later was it discovered that it was calcium and magnesium compounds dissolved in the water that were responsible for the greater or lesser soap requirement. The terms "hard" and "soft" derive from the sensation evoked by soap lather on the skin in the water in question.

From an aquarium viewpoint it is mainly the so-called "carbonate hardness" (KH, expressed in degrees) that is important. It is a measure of the compounds calcium and magnesium carbonate, which react with carbonic acid to form calcium and magnesium bicarbonate. Because they are chemically unstable, both these substances play an important role in the aquarium. They react reciprocally with carbonic acid and can be problematical per se for so-called "softwater fishes" that practically never encounter them in nature. In addition there are yet other calcium and magnesium compounds in water, which, however, are relatively stable chemically and of no great practical significance. These are designated "non-carbonate hardness". The two forms of hardness combined make up general hardness (GH), also measured in degrees, which in this case vary from country to country - those used in this book are German, °dGH. 0-4 °dGH denotes (roughly) very soft, 4-8 °dGH soft, 8-12 °dGH medium hard, 12-18 °dGH hard, 18-30 °dGH very hard, and more than 30 °dGH extremely hard, water.

The pH value is closely connected with hardness, although they are totally separate concepts chemically. The pH value denotes the degree of acidity of the water. It is important to realize that pH is measured using a logarithmic decimal scale, so that water with a pH of 5 is 10 times as acid as pH 6, and 100 times as acid as pH 7. Because the components of carbonate hardness react very strongly with acids, in the aquarium mainly with carbonic acid, the concepts of hardness and pH are very much intertwined from an aquarium viewpoint. Water with a pH of 7 is designated neutral, water with a ph above 7 is termed alkaline, and that with a pH below 7 is acid. The extremes that (specialized) fishes can tolerate are an acid pH of 3.5 and an alkaline pH of 9.5.

The pH can fluctuate dramatically with the day-night rhythm, and this is often the reason why fishes become sick or die. The reason for this pH fluctuation is that at night plants do not use carbon dioxide as they are not engaged in photosynthesis, and in fact actually give off additional carbon dioxide via their respiration. In hard water this has little effect, as the carbonate hardness "cancels out" the carbon dioxide (the technical term for this is "buffering"). However, soft water has little or no buffering capacity (i.e. carbonate hardness) and this can result in pH surges that are life-threatening for the fishes.

There are three methods of avoiding this danger. Firstly an airstone can be used in the aquarium at night. Carbon dioxide is highly volatile and can thus easily be driven off from the water. Alternatively, humic acid can be added via peat filtration or as a liquid preparation, and this will also have a buffering effect, though this method can be used only for fishes that will tolerate acid water. Otherwise, for fish that don't like acidity, the water must be artificially hardened - method 3.

Hardness and pH should be monitored regularly.

# Feeding Corydoras

Mailed catfishes are very easy to feed in the aquarium, as they will take practically any of the usual aquarium fish foods. The various flake, tablet, and granular foods, frozen foods (mosquito larvae, Artemia, Cyclops, fish roe, plus the very good mixed foods), and live foods (mosquito larvae, Daphnia, Tubifex, whiteworms, etc). Nevertheless many go hungry in the aquarium - especially the small specimens. Why is this? Well, first of all, mailed catfishes are not just scavengers or garbage pails. Just like other aquarium fishes they need to be fed carefully and according to the needs of their species. It is rare for the food left by the other aquarium residents to be sufficient for their needs.

The problem of too coarse a substrate has already been mentioned (pp. 11-12).

Another problem, not adequately appreciated by many aquarists, is that mailed catfishes are continuous feeders. Their body is not adapted to ingesting large amounts of food at one go. As many small portions as possible at intervals throughout the day will suit them very well. Of course, not everyone can feed several times per day, but tablet food, which dissolves slowly, is an ideal solution to providing a constant food supply throughout the day. Then, in the evening, they can be given a good portion of live or frozen food. But always make sure that the mailed catfishes get their share and the whole lot isn't eaten by their tankmates!

This is the only problem with housing these catfishes with other fish species. They are without exception peaceful towards other fishes. Thus greedy feeders, for example many barbs, can make it very difficult to ensure the catfishes get enough. The result if often overweight barbs and emaciated catfishes. I personally recommend small characins or cyprinids (Rasbora species and their allies) and livebearing toothcarps as ideal tankmates.

Serious mistakes are often made with flake food (the same applies to tablets and granules, of course) - by the aquarist, not the manufacturer. It is fair to say that any of the flake foods on the market is of suitable quality. But, in order for the food to retain its nutritional value, care must be taken with its storage in the container. It must be kept cool, dry, and in the dark. Some of the ingredients of flake food, absolutely essential for fish nutrition, are easily spoiled – vitamins, unsaturated fatty acids, and other materials are adversely affected by oxygen, light, and moisture. Once the container has been opened the contents should be used within two weeks. Hence buying cheaply in bulk is a false economy! If you use an automatic feeder, it should be refilled with fresh food every day. Feed only as much as is eaten completely within five minutes.

*Corydoras araguaiaensis.*

# Disease in mailed catfishes

Like all life forms, mailed catfishes can suffer from a large variety of diseases. It is sensible to have available a number of medications, for the commonest fish diseases, so that treatment can be rapid in the event of infection.

The onset of disease is almost always evident from a striking change in behavior. If a mailed catfish lies listlessly in a corner of the aquarium, swims with a noticeable wobbling motion, clamps its fins, eats little and without enthusiasm, and they eyes are sunken deep in their sockets, then these are always serious warning signs. Often at this early stage of the disease it is still possible to stimulate the natural immune response of the fish by performing a large water change and raising the normal maintenance temperature by 3-4 °C. Very many fish parasites die or are at least severely weakened if the temperature is somewhat higher than 30 °C. But note: at such high temperatures it is essential to observe the fish very closely. An airstone must always be placed in the aquarium when the temperature is raised to this degree. The oxygen-absorption capacity of water decreases as temperature rises. Paradoxically such high temperatures can also lead to oxygen toxicity through an excess of oxygen in the water if the aquarium is densely planted. At higher temperatures the plants will photosynthesize at a higher rate. An airstone will help in either eventuality. Of course, unlike most other fishes, our mailed catfishes can fall back on their auxiliary gut-breathing. But this will stress them severely, and our patients need all their strength to fight the disease.

Because sick fishes eat very little, they should be fed only sparingly, if at all, in order to avoid additional harm to the patients through poor water quality.

The most important method of treatment of fish diseases is prevention! This includes, of course, not just regular maintenance of the aquarium but also the quarantining of new arrivals. Never introduce a newly-purchased fish into an established fish population immediately. Every fish carries pathogens. This is not the fault of the breeder or dealer, it is simply the way things are. Netting, transportation, and acclimatization to different water conditions are all stressful, with negative effects on the fish. This can lead to an outbreak of disease that the fish would previously have fought off easily. On the other hand, the fishes already swimming in your aquarium will also be carrying pathogens. The new arrival, in its weakened state, can very easily become infected and even seriously ill. The only sure way to avoid these dangers is a small extra aquarium, a so-called quarantine tank. The quarantine tank for mailed catfishes does not as a rule need to be particularly large. Because one should always purchase 10-12 specimens of any new species, then, based on the space requirements of these fishes, a quarantine tank of 60 x 40 cm bottom area should be used.

The quarantine tank should be filled with water from the main aquarium. A heater-stat and a small internal filter will complete the set-up. Don't bother with substrate or plants, which will only be a nuisance in a quarantine tank. However, unlike that for most other species, the mailed catfish quarantine tank should always have a 2-3 cm deep layer of fine sand on the bottom.

No prophylactic medication should be used in the quarantine tank. With luck the new fish will recover very rapidly from the stress and never fall ill. All that is required is to monitor the water parameters in the quarantine tank regularly. If, after two weeks, the fish shows no obvious external signs of disease,

# Disease in mailed catfishes

This *Corydoras cf. leopardus* is really sick. The reddish shine on the belly and the destroyed borders of the fins indicate a bacterial infection. Additionally, this specimen is badly emaciated. It is quite unlikely that this fish will recover again.

# Disease in mailed catfishes

then it can be transferred to the main aquarium.

If treatment is necessary, then again the quarantine tank offers many advantages. Firstly, much less medication is required in a small tank. Secondly, the water change required after treatment is quick and easy to carry out. Thirdly, the treatment can be administered more precisely, as substrate, plants, and also a large filter can considerably hasten the breakdown of the active ingredients of the medication. And fourthly, the undesirable side-effects of some medications (dead snails, damaged plants, intolerance of some active ingredients by some fish species) are irrelevant in the quarantine tank.

Mailed catfishes tolerate some common ingredients of fish medications very badly. Never use remedies containing copper, or the chemical Trichlorfon (contained in, for example, Neguvon and Masoten). Essentially, care is required with medications used to treat worm infestations in fish. Mailed catfishes as good as never suffer from worms. If the other fishes require treatment for worms, then net out the mailed catfishes and put them in another aquarium during the treatment. Another dangerous chemical is malachite green oxalate, a component of very many fish medications. Follow the dosage instructions to the letter! In my experience mailed catfishes will then come to no harm, but an overdose is likely to be fatal!

It is not possible to list here all the diseases to which mailed catfishes may succumb, but there are fat volumes available, devoted to fish diseases. There are, however, a number of particularly common ailments of which every aquarist should be aware.

First there is the notorious white spot disease (Ich).

If the catfish exhibits white, pinhead dots on the body and fins, this in all probability this will be Ichthyophthirius. There are many effective treatments for this pathogen available in the aquarium trade. The treatment will be significantly enhanced by raising the temperature. The reproduction of this parasite involves a free-swimming stage, large numbers of which are released into the water in search of new victims. Elimination of the free-swimming stage is the basis of treatment. Because the free-swimmers always attack bottom-dwelling fishes first, mailed catfishes give an early warning of this disease, and the aquarist can then start treatment while the other fishes in the aquarium show little or no signs of illness.

Similar symptoms characterize "velvet", a second common parasitic disease, but this time the dots are very small. This is Piscinoodinium, better known as Oodinium. Again, numerous treatments are available, but beware – some of them contain copper (the composition of a medication should be printed on the container). Because the parasite often initially attacks the gills, rapid respiration coupled with apathetic behavior are early symptoms. But note, nitrite poisoning causes similar symptoms! So a nitrite test should be carried out before treatment.

Bacterial diseases are dreaded and, unfortunately, difficult to treat, but commonly occur in mailed catfishes if they are imported during the dry season when they have already had a close brush with death and are correspondingly badly stressed. In mailed catfishes bacterial diseases typically manifest as red spots ("red blotch disease"), ie areas of subcutaneous haemorrhage. The only effective treatment is antibiotics, which should be administered only under the supervision of a vet. Treatment with antibiotics is not without dangers

# Disease in mailed catfishes

for the aquarist, and should never – and this is no idle warning – be undertaken "off one's own bat". Bacterial diseases can also be combated without chemicals by optimizing living conditions: A1 water quality, high-protein food (especially Tubifex) and warmth will improve the immune response of the fish and often prove successful. Bacterial diseases are generally the result of earlier unfavorable living conditions, developing only when the fish has been weakened by other factors. The causative pathogens in such cases are often normally harmless bacteria responsible for the breakdown of dead plant and animal material. Reduction of the micro-organism population in the water via a UV sterilizer is thus a good method of prophylaxis.

In the event of a bacterial infection, always look for the root cause. Has the pH been fluctuating with the day-night rhythm? Is there a troublemaker in the aquarium, continually causing disturbance and stress? Could there be an unsuspected parasitic infestation? Is the filter properly maintained, and has a partial water change been performed every week? Is the tank overpopulated?

Badly emaciated fishes can be fattened up again only with much care. They need "baby food" in the form of live Artemia nauplii (even if they are large fishes) and plenty of peace and quiet. Don't give them normal foods (especially not high protein foods such as granules or Tubifex) until it is obvious that they are on the road to recovery.

I must also mention a peculiarity of mailed catfishes - they are able to poison themselves. I once planned to give a friend 10 mailed catfishes for his large aquarium, as a birthday present. They were splendid, full-grown Corydoras pulcher. I packed them in ample containers and they were at most two hours in transit thereafter. But when they arrived at my friend's they were all belly-up, half dead. I was absolutely stunned. The transport water was cloudy and foaming! Happily they all recovered when placed in clean water. To date little is known about this phenomenon. But it appears that some species at least (the same has been reported to me regarding Corydoras sterbai) are able to produce a kind of poison via their skin, which in the wild undoubtedly serves to repel predators. In the confines of a plastic bag this secretion can lead to self-poisoning. So far this has been reported only from species with bright, light-colored spines in the dorsal and pectoral fins - and "spiking" by these is also rather painful. Obviously, when these species are to be transported for any length of time, the water in the bags or other transportation containers should be changed completely an hour after the fishes were netted. By then they will have settled down again.

Under certain circumstances Corydoras pulcher can poison themselves.

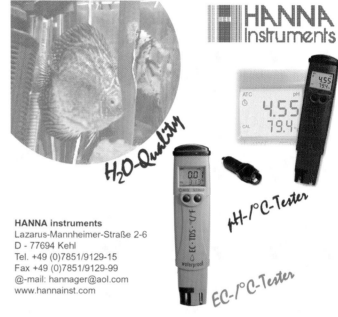

# Copy-cats!

One fascinating phenomenon among mailed catfishes is that for very many species there are species pairs or species trios that have (almost) identical body markings but completely different head shapes, and which moreover are not closely related. These forms are termed blunt-snouts, saddle-snouts, and long-snouts.

The three groups exhibit different feeding behaviors. Blunt-snouts feed mainly on food particles lying on top of the sand. Saddle-snouts usually burrow a few millimeters into the sand, thereby reaching any food particles buried somewhat deeper. And long-snouts dig into the sand up to their eyes and thus reach yet other food particles.

This behavior appears to be accompanied by other anatomical adaptations. Thus blunt- and saddle-snouts spit out the sand they pick up (which is no use to them) via their mouths. By contrast the deep-digging long-snouts expel unwanted sand via their gill-openings.

At some time in the future these three ecological groups (as well as a number of other, readily distinguishable, species groups) will probably be transferred into separate genera. They are certainly not closely related and are not monophyletic in their origins (monophyletic means that all species in a group originally derive from a single ancestral species). However, that is not our concern here. Much more important is the question, why do these species imitate one another in their markings?

It is a basic precept of ecology that similar species can live together in the same habitat only if they occupy different ecological niches (in this case, different food resources). Otherwise the level of competition between them would be so high that only a single species could survive. That explains why three species can live together, but what is the advantage of the similar coloration? Ultimately *Corydoras* have as good as no enemies as their defensive tactics (bony armor and fin spines) are exceptionally effective. To date there is no satisfactory answer to these questions. But perhaps we are simply asking the wrong questions. Perhaps we should be asking, what is the general purpose of color patterns in *Corydoras*? Why aren't all species simply the same color? Perhaps the color pattern that a mailed catfish species exhibits in a particular habitat is not just striking and attractive, but represents the optimal (for reasons not apparent to us) coloration for that particular habitat. And so perhaps the blunt-, long-, and saddle-snouted forms do not imitate each other at all, but are instead simply all optimally adapted to their habitat. Who knows?

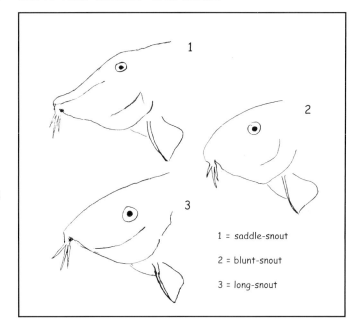

1 = saddle-snout

2 = blunt-snout

3 = long-snout

## Species trio 1: blunt-snout

*Corydoras arcuatus* is a very popular species, regularly available. It grows to about 6 cm long.

## Species trio 1: long-snout

The long-nose analogue of **Corydoras arcuatus** is not yet scientifically described. It is known as "Narcissus II" in the trade. At about 8 cm long, it is significantly longer than the round-nose.

## Species trio 1: saddle-snout

The saddle-nose analogue of *Corydoras arcuatus* is called *Corydoras narcissus*. The species name refers to the youth Narcissus, who fell in love with his own reflection. The fish was given to scientists for study with a note to the effect that it "must" be named in honor of a particular person. Han NIJSSEN and Isaäc ISBRÜCKER were less than impressed and, tongue in cheek, named the approximately 8 cm long fish after the self-centered mythological figure!

## Species trio 1: What's going on here?

Often there are more than three species with the same markings. This scientifically undescribed species is called "Super Arcuatus" in the trade. It looks like a **Corydoras arcuatus** but is twice the size: 12-15 cm. Perhaps these giants inhabit deeper parts of rivers while their smaller cousins live in the shallows.

## Species trio 2: blunt-snout

*Corydoras adolfoi* is a gorgeous, up to 6 cm long, mailed catfish with a particularly bright orange spot in front of the dorsal fin.

## Species trio 2: long-snout

***Corydoras imitator*** is the long-snout analogue of *Corydoras adolfoi*. It occurs in the same habitat, but, like most long-snouts, grows somewhat larger than the blunt-snouted species, in this case about 8 cm.

## Species trio 2: saddle-snout

***Corydoras serratus*** is the saddle-snout analogue of *Corydoras adolfoi*. This splendid, about 8 cm long, fish hardly ever reaches Europe, as the small number of individuals exported fetch incredibly high prices in Japan, where *Corydoras* mania is all the rage.

## Species trio 2: What's going on here?

***Corydoras nijsseni*** is a 4-5 cm species of the *Corydoras elegans* species group. Like all these fishes, it is noted for sexual dichromatism (different coloration in the two sexes) and because it prefers to swim in mid-water.

# Shoaling fishes?

One of the generalizations about mailed catfishes that we hear time and again is that they are peaceful shoaling fishes. Is that true? In my opinion, no!

Although it is true that it is extremely rare for any other fish in the aquarium to be harmed by a mailed catfish (unless it was trying to eat the catfish and choked), it is only because these catfishes do not have any offensive weapons, and hence are incapable of harming other fishes, that they have long been regarded as peaceful. But the members of some species groups are very prone to squabbling among themselves and the result long-term stress can even lead to losses. Limited population levels in the aquarium, and adequately large tanks, are a sure way of preventing such problems.

And as for "shoaling fishes" – I would prefer to describe the majority of mailed catfishes as pack-forming. There is no doubt that all species have well-developed social behavior and do not do particularly well when kept single. But this does not make then true shoaling fishes. As a general rule of thumb, the smaller the fishes, the larger the group should be. Thus the tiny *Aspidoras pauciradiatus*, *Corydoras pygmaeus*, and *C. hastatus* should be kept in groups of 20-30 individuals, the other, medium-size *Aspidoras*, *Corydoras elegans*, *C. napoensis*, *C. nijsseni*, *C. breei*, and *C. osteocarus* – in short, all species that attain a length of 4-5 cm – in groups of 10-15 individuals, and all the rest in groups of 5-10.

Of course it is possible, given aquaria of sufficient size, to keep more individuals per species – that should be self-evident. Some long- and saddle-snout species, however, appreciate their own private space. In such cases allow 15-20 cm2 of bottom area per individual.

Although not strictly relevant here, it is worth noting that many of the spotted species in the 5-7 cm range are regularly found en masse in large shoals in the wild. That is also the reason why these species are always available in the trade and at a very good price.
The popular "leopard corys", *Corydoras julii* and *C. trilineatus*, belong to this group, as do, for example, *C. ambiacus* and *C. agassizii*.

*Corydoras* sp. aff. *sanchesi*

# More mimicry

When the coloration of one species resembles that of another, this is termed mimicry. The best-known examples of mimicry are those where totally harmless creatures imitate genuinely poisonous ones and profit from the bad reputation of their models. Among the insects, for example, there are the totally weaponless hoverflies that imitate the stinging wasps, bees, and hornets.

There are species of naked catfishes (specifically *Brachyrhamdia*) that are found among mass imports of particular *Corydoras* species on a puzzlingly frequent basis. Sometimes these "pim" species bear a close resemblance to the corys in their coloration. But this mimicry is not always apparent at first glance. *Brachyrhamdia* that are isolated for photography sometimes look very different to the *Corydoras*, but they are very difficult to pick out among the shoal. Anyone who has ever tried to net out such "by-catches" from a tank containing perhaps 800 mailed catfishes will be well aware that the task is enough to drive you out of your mind.

In this case the mimicry is much easier to explain than that which exists between mailed catfish species. Compared with mailed catfishes, the *Brachyrhamdia* species are quite defenseless, even though they have strong, dagger-sharp, pectoral-fin spines. It is mainly young specimens that are found among mailed catfishes, where they apparently find protection among the throng of corys. These *Brachyrhamdia* species rarely exceed 10 cm in length, and I can only suggest you snap them up whenever you have the chance. There is undoubtedly much to be learned about this interrelationship by keeping them together and observing them carefully. For example, it may be that the pims aren't in fact seeking shelter among the corys at all, but are instead interested in the small organisms stirred by the digging throng, and which are the main prey of these pimelodid catfishes.

*Brachyrhamdia marthae* is one of the more regular imported species of the genus. As in all *Brachyrhamdia* species, the males develope enlarged fins.

# The Paraguay connection

The final mimicry to be discussed here in detail is practiced by the dainty sickle-spot cory, *Corydoras hastatus*. With a maximum length of 3 cm, this is one of the smallest *Corydoras* species known. Like the even smaller pygmy cory, *Corydoras pygmaeus*, *C. hastatus* almost always swims in open water and has largely abandoned the bottom-dwelling life. *C. hastatus* is found mainly in the Rio Paraguay and its tributaries which rise in the highlands of Mato Grosso in Brazil and then flow through Paraguay.

Here *C. hastatus* mingles with the very similarly colored characins of the species *Cheirodon kriegi*, *Aphyocharax paraguayensis*, *Psellogrammus kennedyi*, and *Hemigrammus tridens* to form mixed shoals. The advantage of this mimicry is obvious: the dwarfing of this catfish has led to an increase in the number of species that are no longer frightened off by its feeble armor. So it is now subject to the same rules of the game as all other small defenseless fishes that have found the same answer to this problem: strength in numbers. A typical feature of all the members of this mixed, symbiotic (symbiosis = the living together, for mutual benefit, of different species) shoal is a black, light-ringed spot on the caudal peduncle, intended to look like an eye to any predators. When attacking, predator fish orient themselves on the eye of their victim and so the presence of an ocellus (= eyespot) on the posterior body increases the likelihood of the predator grabbing "thin air". This trick has been discovered over and over again by a huge variety of fish species.

*Corydoras hastatus*, the sickle-spot cory.

The mixed Paraguay shoals have a great advantage over single-species shoals: each member species has a different food requirement and so there is far less competition among the individual members of the shoal than would be the case in a single-species shoal. Thus, if a shoal consists of, say, 1000 individuals, then a single *Aphyocharax paraguayensis* will have to compete with 999 conspecifics for a single fly (its main prey) that falls into the water. In a mixed shoal of five species, on the other hand, it will have only 199 competitors, and yet the protection factor remains the same.

Unfortunately the above-mentioned *Aphyocharax paragu-*

# The Paraguay connection

*ayensis* is the only one of the species mentioned earlier that is imported regularly - the others are too colorless. So only a few lucky people have had the opportunity to keep all these species together in a large community aquarium and to study all the nuances of this symbiosis. It is a shame that it is invariably just the colorful fishes that sell well in the aquarium trade

*Aphyocharax paraguayensis*

*Cheirodon kriegi*

*Hemigrammus tridens*

# Floating eggs – the *C. elegans* group

On the following pages the individual mailed catfish species groups will be presented and discussed. Because the differentiation of the sexes and the special breeding requirements of these species groups are rather different, these will be discussed for each group in turn.

However, I will limit myself to the species that are regularly available in the aquarium trade as, in the final analysis, this is not a complete scientific treatise on the genus *Corydoras*.

The species of the *Corydoras elegans* group are rather uniform in their appearance. It is easier to assign these fishes to their species group than to individual species. At present, seven rather variable species are recognized: *C. elegans*, *C. napoensis*, *C. nijsseni* (photo on p. 37), *C. undulatus*, *C. latus*, *C. bolivianus* and *C. nanus*. Whether there are really only six species, or whether the at least 15 regularly imported "color variants" will later be designated separate species, who knows? One feature they all have in common is an elongate oval body shape, with a small head and almost terminal mouth. In addition, all species exhibit clear sexual dichromatism: the males are noticeably more colorful and more heavily patterned than the females, and also remain appreciably smaller. All species have comparatively small eggs and fry. They are sometimes less than cooperative when it comes to breeding. The easiest to breed is *C. napoensis*, followed by *C. nijsseni*. Provided the fishes are in breeding condition, the other species can sometimes be induced to spawn by moving the entire group into another aquarium.

All species are very attractive and fairly problem-free aquarium fishes that can be recommended for beginners as well.

The males like to squabble among themselves. In fishes of this group there is an observable tendency to swim in midwater, abandoning the strict bottom-dwelling life-style.

One interesting species, closely related to this group but apparently rather specialized, is the large-growing *C. pantanalensis*, often known as "C 5" in the aquarium hobby. This colorful, about 10 cm long, species exhibits sexual dichromatism only at breeding time, when the sexes look so different that they could easily be taken for two separate species!

*Corydoras pantanalensis*, ♂

*Corydoras pantanalensis*, ♀

## Floating eggs – the *C. elegans* group

*Corydoras napoensis* ♂

*Corydoras napoensis* ♀

*Corydoras* sp. "San Juan", ♂

*Corydoras* sp. "San Juan", ♀

*Corydoras elegans,* ♂

*Corydoras elegans,* ♀

## Mini-corys

The pygmy cory, *Corydoras pygmaeus*, and *Corydoras hastatus* deviate significantly from the mailed catfish norm in their lifestyle. This has already been mentioned in the chapter "The Paraguay connection".
Their body shape is highly reminiscent of the *C. elegans* group, from which they differ in that males and females exhibit the same coloration. However, among the mini-corys too, the males remain smaller than the females.

Both species are fairly easy to breed provided they are allowed to become well-established in a well-planted species aquarium (i.e. without any other fish species). If the adults are fed morning and evening each day on Artemia nauplii, and if a large partial water change is performed weekly, then fry will appear almost automatically, and a shoal of some 30 individuals can then prove nicely productive.

A second group of mini-corys is represented by *Corydoras habrosus*. Another, similar species, but rarely available in the trade, is *Corydoras cochui*. Although this group of mini-corys barely exceed 3.5 cm in length, they are, however, typical bottom-dwelling fishes, and not closely related to the *C. elegans* group. It is my belief that they are miniaturized relatives of the *C. paleatus* group. Because of its attractive black-white pattern, *C. habrosus* is also known as the checkerboard catfish. The breeding of this species presents no great problems, and normally occurs after a large water change using cool water, which simulates rainfall in the wild.

*Corydoras pygmaeus*

*Corydoras habrosus*

*Corydoras cochui*

# Hatching Artemia

*Artemia salina* belongs to an ancient group of crustaceans, the so-called branchiopods (gill-feet). They are characterized by all species having adapted to periodic drying-up of their habitat by producing reproductive cysts, often termed eggs by aquarists. These "eggs" can survive in the bottom mud for weeks, months, even years of drought. Brine shrimps have adapted to a particularly saline habitat, although other branchiopods react rather badly to salt. Naturally, in their extreme habitat *Artemia* have no enemies and can proliferate massively – which is the basis of this small creature being used commercially as a fish food. The Artemia available in the trade originate mainly from the great salt lakes of the USA. Adult Artemia are about 1.5 cm long.

While most other food organisms live in waters also inhabited by fishes, this is not the case with *Artemia*. Hence brine shrimp are never carriers of fish diseases, *Artemia* is an indispensable food during the acclimatization of stress-sensitive wild-caught fishes.

*Artemia* eggs can be purchased at any aquarium store. Salt water is required for hatching, and the salt concentration should be between 3 and 8% (= 30-80 g/l). The simplest way is to add 3 rounded tablespoonsful of salt to a liter of water if nauplii are required quickly because unexpected breeding has taken place. Some nauplii will always hatch at this salinity. But for an optimal hatch rate, if time permits, it is necessary to experiment a little with the salinity. Very often the salinity required for an optimal hatch rate will vary somewhat from batch to batch of eggs. At a temperature of 18-32 °C the nauplii hatch after 24-36 hours. The hatching time is temperature dependent. The culture should be left to stand for 48 hours before harvesting to obtain the maximum "crop".

It is easiest to use ordinary household cooking salt for culturing *Artemia*. But you must always make sure that the cooking salt doesn't contain added fluorine or iodine salts, as the *Artemia* will not tolerate these. Cooking salt sometimes has additives to make it flow freely – these will not harm the *Artemia*, but they will affect the strength of the salt solution. Often by the end of the packet there is more flow agent than salt left. I therefore recommend coarse crystal cooking salt, as sold for salt mills. This contains no flow agents and is easy to measure. The use of salt intended for marine aquaria is a luxury, and this rather expensive salt does produce an exceptional hatch rate. It should above all be used if you want to rear the *Artemia* and feed them to larger fishes.

To hatch a large quantity of *Artemia* eggs (up to about half a tablespoonful) you will need an airpump, airline, and an empty 1 liter bottle (clear wine bottles have proved very good) or a manufactured *Artemia* hatchery. For continuous feeding it is best to use two bottles (cheers!) or two hatcheries. It is all very easy, except that some people may find the bubbling of the bottles and the humming of the airpump irritating. If only small amounts of brine shrimp are required (for about 30-50 fry) then there is no noise. In that case you need only small (300 ml) jam jars filled with salt solution, and a knife-tip-full of eggs sprinkled on the surface. The surface tension of the water will keep the eggs at the surface until they hatch, so they will have plenty of oxygen.

*Artemia* nauplii always swim towards the light (this behavior is termed positively photo-active). To harvest them, turn off the aeration and tilt the container slightly towards a strong light source. The nauplii will then congregate on the side nearest the light, while the most recently hatched, still very young, *Artemia* congregate at the bottom of the container. Now they can simply be siphoned off with airline into an *Artemia* sieve. Don't forget to turn the aeration back on!

## Cool sideburns

To my eye the most attractive, but unfortunately – under normal aquarium conditions – most delicate mailed catfishes are *Corydoras barbatus* (barbatus = bearded) and its relatives. These fishes all display visible sexual differences. The males have much brighter colors, significantly larger fins, and, when sexually ripe, bristly "sideburns". This group includes the species *C. barbatus* and *C. macropterus*, as well a number of rarely imported species and – in my view – also the unfortunately rarely available *Aspidoras virgulatus* that is in many respects closer to the *C. barbatus*-group than to other members of his genus. They are all very attractive species. *C. barbatus* can grow up to 12 cm long, *C. macropterus* to about 8 cm, while *Aspidoras virgulatus*, at about 4.5 cm, is the smallest member of the group.

Only *C. barbatus*, which has the common names "bearded cory" and "checkerboard cory", is regularly available in the aquarium trade. Unfortunately most buyers are unaware that this species is found far to the south and in the highlands and hence requires cooler temperatures. Breeding is most likely and the fertilization rate higher at 18-20 °C. During the spectacular courtship the males sail around their females with fins wide-spread. The rudiments of brood care have been observed in bearded catfishes, with males sometimes guarding the spawning site. Unlike many other *Corydoras* species, which lay their eggs in batches all over the tank, *C. barbatus* have a tendency to lay all their eggs in the same place.

If species adapted to cool water are kept too warm in the long term, then first they will stop breeding, and this is followed by permanent emaciation and eventual continuous ill-health. Hence anyone who is interested in these splendid creatures should set up a special aquarium maintained at the correct temperature. The aquarium for *C. barbatus* should not be too small – the bottom area should be at least 120 x 50 cm.

*Corydoras macropterus*, ♂

*Corydoras macropterus*, ♀

*Corydoras barbatus*, ♀

# Cool sideburns

Male *Corydoras barbatus*, the checkerboard cory, are the most beautiful mailed catfishes of all.

# Gleaming metal

Among the best-known mailed catfishes, always available in the aquarium trade, easy to keep and breed and very attractive with it, are the members of the *Corydoras aeneus* species group, the metallic mailed catfishes.

The majority of species lack any distinctive markings, but instead their bodies have a metallic sheen in a huge variety of shades of color. Some are warm gold-bronze, some are metallic green, some are coppery-red – there is hardly a metallic hue that is not found in one species or another. The best-known species of this group is undoubtedly *Corydoras aeneus*, commonly known as the bronze catfish. But perhaps bronze catfishes would be a better name, as there is increasing evidence that what at present is regarded as a highly variable species with a huge distribution is in fact half a dozen or more different species. During cell division, the genetic information in the cell, the DNA, becomes highly concentrated and can be made visible in the form of the so-called chromosomes. The number and expression of the chromosomes is very characteristic for each individual species. Chromosome studies of bronze catfishes have shown that individuals from different distribution areas, despite closely resembling each other in external appearance, are quite different in their chromosome counts and make-up. Hence it is perhaps better to speak of the *Corydoras aeneus* species complex.

Other species of the group, which have a very similar, rather elongate body shape and are typical bottom-dwelling mailed catfishes, are the attractive *C. melanotaenia* and two species conspicuous by their dark dorsal stripes, *C. rabauti* and *C. zygatus*. This group too contains further, but practically never imported, species. Many of the metallic catfishes are so easy to breed successfully that they are most heartily recommended to the beginner. Often spawning will take place all year, even in the community aquarium. In this event, the highly adhesive spawn, which is attached all over the tank in batches of 10-25 eggs (the aquarium glass is a favorite site) can be carefully removed with a razorblade and hatched in a small spare tank. Large females can produce up to 500 eggs! *Corydoras aeneus* grows to about 6 cm long, *C. zygatus* is the largest member of the group at about 7 cm.

Males remain smaller and slimmer than females. The sexes can readily be distinguished from above as males have visibly longer pectoral fin spines than females.

*Corydoras melanotaenia*

*Corydoras rabauti*

*Corydoras* cf. *aeneus* ( = "*C. schulzei*")

# Gleaming metal

This fish, which is very similar to the bronze catfish, is sometimes imported from Peru. Its coloration is splendid.

# Round heads and high backs

All the species in this group – *C. breei, C. armatus, C. loretoensis, C. osteoparus, C. sipaliwini* (syn: *C. bondi bondi*), *C. reynoldsi* and *C. xinguensis* are fairly regularly imported – share the common feature that they look as if they have run full tilt into a brick wall. The head is relatively short, while the back rises steeply upwards, and the overall impression is reinforced further by the often greatly elongated dorsal fin spine. All species remain rather small (about 4-5 cm long) and hence are ideal aquarium fishes. Unfortunately they are often severely weakened by transportation and hence require careful acclimatization.

At breeding time in some species a striking hopping behavior, by the male in front of the female, can be observed. Unfortunately the species are not generally easy to breed, and there have been no reports of success at all for some of them. Even the attractive and much-prized *C. reynoldsi*, whose first importation a few years ago caused a sensation among mailed catfish fans, and which was first bred relatively quickly, has not yet been included in professional breeding programs: too few eggs per spawning, and too much effort required in rearing make this species commercially uninteresting. In general, males of this group remain smaller and slimmer than females. It appears that in the wild they spawn over a long period, with pairs briefly separating from the rest of the throng and laying a small number of eggs. Individuals not involved in spawning appear to predate heavily on the eggs. The breeding of one species (*C. breei*) was first achieved in a completely darkened aquarium with only a viewing slit left uncovered. Perhaps these species inhabit heavily shaded forest streams in the wild and are unsettled by too much light.

*Corydoras loretoensis*

*Corydoras armatus*

*Corydoras sipaliwini*

# Round heads and high backs

The first importation of *Corydoras reynoldsi* a few years ago caused a sensation. Unfortunately this species cannot be bred "efficiently" and hence is only irregularly available in the trade.

# Black and white pandas

A whole series of mailed catfish species exhibit a small number of striking black markings on a whitish or sand-colored background. That may sound boring, but it isn't! Hence the members of this group of related species are perpetual favorites in the aquarium trade, and constantly represented by one species or another. Important aquarium fishes in this group are *Corydoras panda*, *C. metae*, *C. melini*, and *C. davidsandsi*. Other species, such as *C. oiapoquensis*, etc, are only very rarely available.

The breeding of *Corydoras panda*, *C. metae*, and *C. davidsandsi* is very easy and can be achieved without problem even by beginners. These fishes will even often start spawning in the community aquarium. The number of eggs per female is generally about 30, but it has also been reported that *C. davidsandsi* can produce up to 80 eggs per female per spawning. The sexes are not very easy to differentiate, especially in younger specimens. In general, males are somewhat smaller. It helps to examine the fishes from above (in a bucket, for example), as males have longer pectoral fin spines, even at a relatively early age. It often happens that young *C. metae* develop a large, almost sail-like dorsal fin, but unfortunately this disappears as they grow larger. Nothing is known of the significance of this phenomenon. The colors of these sandy bottom dwellers can be seen to best effect if they are kept in a well-lit aquarium with a light-colored substrate. Otherwise the body color becomes much darker and they lose their former glory. From a behavioral viewpoint, these approximately 5 cm long fishes are very much classic mailed catfishes.

*Corydoras davidsandsi*

*Corydoras panda*

# Black and white pandas

*Corydoras metae*, named after its home, the Rio Meta in Colombia, is sometimes labeled "Bandit" on wholesaler price lists, and the dark eye stripe does indeed give these charming fishes a rather rakish look!

# Evolving sand-dwellers

A further group of light-colored mailed catfishes are very similar to the pandas, and in my opinion also form a natural unit. It is always rather difficult to establish which members of a species group are ancient and which more recent. But it is my belief that, because of their enormous observable intraspecific variability, the species discussed here are sill in the process of evolving, i.e. they offer us an opportunity to see evolution in action. For this reason the differentiation of the species listed below is not easy, very often there are intermediates. Species frequently offered for sale include *C. axelrodi*, *C. loxozonus*, and the as yet undescribed species "C 3". All of them are also traded under the scientifically meaningless name "*Corydoras deckeri*". There are also other species in this group, but they are only very rarely imported.

As already hinted, there are numerous "color varieties" of these species and it is as yet completely unclear whether these are already separate species or not. In practical terms this signifies that it is essential always to breed from pure strains. Better to let a strain die out and instead obtain a new breeding group with a shared origin than to produce an aquarium form by buying extra or replacement individuals. There are so many natural species of mailed catfishes that we really don't need any "man-made" forms.

Diverse "variants" of *C. axelrodi* are constantly available in the aquarium trade. This species breeds easily and is productive. In general, about 30-50 eggs per female per spawning can be expected. Like *C. metae*, the species referred to as "C 3" has an enlarged dorsal fin when young.

Sexual differences and maintenance are as for pandas.

*Corydoras* sp. "C3", ♀

*Corydoras* sp. "C3", ♂

# Evolving sand-dwellers

*Corydoras axelrodi* is a "standard" in the aquarium trade because of its beauty and ease of breeding. Its natural habitat is in Colombia, more precisely in the Rio Meta.

# Oddball corys (1)

The regularly maintained mailed catfishes include species that cannot be assigned to any of the various species groups and whose systematic position is hence totally unclear.

They include the beautiful and always available *Corydoras arcuatus* and its as yet scientifically undescribed cousin the "Super Arcuatus". *C. arcuatus* is a fish that is criminally neglected by both scientists and specialist aquarists, because it is perceived as "too commonplace". Only if you are compelled – as I am here – to summarize what is known about these "everyday fish" does it become evident that the answer is, "not much".

The maintenance of these approximately 6 cm long fishes is easy and suitable even for beginners. By contrast, breeding often falls foul of the difficulty in differentiating the sexes. The best method is via the form of the ventral (pelvic) fins: those of females form a kind of pouch in which they carry their eggs for a short time after spawning, until they are attached to a substrate. For the reasons why females have differently shaped ventrals to males, please see the chapter on breeding (p. 68) later in this book.

I have, however, owned wild-caught specimens of a very similar species (perhaps just a geographical variant, but no-one knows) in which the sexes were very easy to recognize: males were visibly slimmer and had a larger and more pointed dorsal fin.

*Corydoras arcuatus* falls somewhat outside the normal framework of the genus. The home of this fish is in Peru, Ecuador, and parts of Brazil.

# Oddball corys (2)

In my view two other species have also taken a separate evolutionary path, although they might be regarded as belonging to the metallic catfish group in its broadest sense: *Corydoras nattereri* and *Corydoras baderi*. Both species have a longitudinal stripe along the center of the flank. *C. nattereri* always has a dark spot in front of the dorsal fin, absent in *C. baderi*. While *C. nattereri* is almost always available in the aquarium trade, *C. baderi* has only been brought back in small numbers by private expeditions.

Sexual differences are as in the metallic catfishes. Interestingly females invariably lay just one egg at a time, producing up to 50 per spawning. Simulation of a rainy period (see "Breeding and rearing") and the presence of numerous ripe females is the key to breeding *C. nattereri*. A further pair of species, again deviating from the norm and commonly available, are *Corydoras concolor* and *C. polystictus*. These too may be descended from metallic catfish ancestors. Both share a common feature: a strikingly rounded and at the same time anterior dorsum (back).

In these species males remain smaller than females. Unfortunately breeding is difficult and belongs to advanced mailed catfish science. Often just keeping them is far from easy as these fishes are inclined rapidly to take exception to errors in maintenance. An increase in the micro-organism level in the water rapidly leads to rotting of the barbels and the death of the fish. Hence these two species are best left to experienced mailed catfish keepers.

*Corydoras polystictus*

*Corydoras nattereri*

# Individualists

A few years ago some gorgeous, about 12 cm long, mailed catfishes were imported from Peru, and until now have remained rarities even in museum collections. The species concerned are *Corydoras semiaquilus* and *C. fowleri*. And here too it is necessary to forget everything you ever knew about mailed catfishes and start again from scratch.

Starting with sexual differences, for example. Normally mailed catfish females are larger than males. Not so in these Peruvians. Here the male is the larger of the two. But the most fascinating phenomenon about these fishes is that each fish has different body markings. Unfortunately so far we still know very little about the natural habitat of these fishes. I think it likely that they live relatively solitary in smaller streams. Given the high degree of genetic variation apparently present in these fishes, they could thus fairly quickly produce a different form in each stream, much like the numerous local forms known among the salmonids. Whether or not it is appropriate to regard these forms as species remains questionable. I have seen importations in which, as well as typical *C. semiaquilus* and *C. fowleri* there were intermediates of every conceivable kind. In the meantime one such local form has been described as *Corydoras coriatae*. I will not delve too deeply into this matter here, as with only a few dozen specimens reaching Europe each year, very few people have been lucky enough to get hold of some. But perhaps you are one of the fortunate ones, in which case your observations can maybe play a part in solving this great puzzle.

The eggs laid by these fishes are rather tiny and hence easily overlooked. As in many rather long-snouted forms a strong current is beneficial if you want to breed them. I myself kept the first *C. fowleri* imported and found them rather easy to maintain. The temperature in my aquaria never exceeds 24 °C. Unfortunately I have no details regarding the long-term maintenance of these fishes.

*Corydoras semiaquilus*

*Corydoras semiaquilus*

## Individualists

**Corydoras fowleri** is a great rarity in the scientific collections of the world.

# Specks of gold

A whole series of mailed catfishes have become very popular aquarium fishes because they have bright yellow or red-gold specks in front of the dorsal fin. There is little doubt that these spots play a role in intraspecific communication, i.e. in the recognition of conspecifics.

Of course, there are gold spots in other species groups as well, for example *C. nijsseni* among the species of the *C. elegans*, and *C. imitator* among the mimics. But those portrayed here are so very similar in general body structure and also behavior, that they probably form a natural unit. Popular species in this group include *Corydoras adolfoi, C. bicolor, C. burgessi, C. delphax, C. schwartzi, C. brevirostris* (syn: *C. melanistius brevirostris*) and *C. surinamensis*. A special characteristic of this group of fishes appears to be a distinct variability in the pattern of markings. Because these fishes are collected together, this must be true variability and nothing to do with local forms. In *C. schwartzi*, for example, the dorsal fin spine can be white, yellowish, or gray. *C. burgessi* there are individuals that have just a single black spot below the dorsal fin, while others have a proper dorsal band, similar to that in *C. adolfoi*. However, *C. burgessi* can always be distinguished from the latter species by its deeper body form. And finally, in *C. adolfoi*, there are individuals with a broad dorsal band and others which exhibit only a narrow streak.

In these species the males remain smaller and daintier than females. In this group too it helps to view the fishes from above in order to differentiate the sexes.
Breeding success with some species (*C. adolfoi, C. burgessi, C. delphax*) has been fairly good, although it is sometimes difficult to induce these fishes to spawn for the first time. However, there are as yet no detailed reports of breeding in the often mass-imported *C. schwartzi*. Perhaps the species is "too commonplace" for anyone to have made any serious attempt?

The species of this group should not be kept too cool – 24-26 °C is an appropriate range. In other respects they are easy to keep and even beginners need have no worries about trying them.

*Corydoras schwartzi*

*Corydoras brevirostris*

*Corydoras burgessi* is a very pretty mailed catfish, regularly available.

# Pet leopards

A whole group of mailed catfishes with a pattern of dots and a black spot in the dorsal fin have been given the rather misleading common name of leopard catfishes. Two species are always available in the trade, two more are uncommon, and a few are as good as never seen: *C. julii*, *C. trilineatus* (these two always), *C. punctatus* and *C. copei* (these two rarely). The true *Corydoras leopardus* (there really is a species of that name) is a long-snout, and not at all closely related to the blunt-snouts listed here.

The leopard catfishes are with justification amongst the most popular mailed catfishes in the hobby. They are attractive, hardy, manageably small (about 4-5 cm) – who could ask for anything more?

Unfortunately the breeding of these fishes is not so simple as one might think from their constant availability in the aquarium trade. This has nothing to do with distinguishing the sexes; that is relatively easy, as males of these species are noticeably smaller and have much longer pectoral fins. The problem is more one of inducing the group to spawn, and this often takes place only after numerous false starts. There are several color varieties of *C. trilineatus*. Interestingly, exactly the same variants occur among the long-snout mimics. Thus, if *C. trilineatus* with an attractive vermiculated pattern on the head are imported, then the *C. leopardus* from the same area will also have vermiculated markings. But if the *C. trilineatus* from another corner of Amazonia have a dotted pattern, then the *C. leopardus* from this region will also have dots on their heads.

The maintenance of the mailed catfishes of this group can be characterized as easy, but they don't like it too cool.

*Corydoras julii*

*Corydoras trilineatus*, ♂♀

*Corydoras punctatus*

# Pet leopards

*Corydoras trilineatus,* the three-line catfish.

# Born mimics

As a rule there is always a long- or saddle-snout form for every blunt-snout species. Sometimes even both. At the same time, many of these saddle-snout mimics appear to be very closely related. Even top mailed catfish specialists might find it difficult to distinguish species as differently colored in life as *C. ellisae* and the scientifically undescribed "Reynoldsi Longnose" on the basis of faded preserved specimens, because there are no significant differences in form. The saddle-snout species *Corydoras blochi*, which is frequently imported, provides a clue to the evolutionary history of these mimics. This species is highly variable in the wild. In a shoal of perhaps 200 specimens there will rarely be two that have exactly the same markings. This is a very unusual phenomenon in the animal kingdom, and it may well be that the mimics can modify their coloration within a relatively short time for an optimal match with whatever blunt-snout species is present. In terms of numbers, the mimics appear always to be greatly in the minority, compared with their blunt-snout cousins. Imports usually contain only one or two saddle-snouts per hundred or more blunt-snouts. In the aquarium the saddle-snouts also exhibit totally different behavior. They are rather territorial and hardly ever swim in shoals.

Among the saddle-snouts, males are significantly smaller than females. These fishes are very delicate and in every case trickier to acclimatize than are many other mailed catfishes. Moreover, it has been shown that the saddle-snouts prefer a rather strong current in the aquarium for breeding. Their eggs are noticeably smaller than those of the majority of blunt-snouts, but at the same time are often more numerous. In *C. ellisae* up to a hundred eggs per female per spawning have been recorded.

By contrast the classic long-snout species are rather uniform in their coloration. They also show no great difference in their behavior to the blunt-snouts. Examples include the species *C. imitator*, *C. leopardus*, *C. maculifer*, and *C. seussi*. Long-snouts are rapid swimmers and can completely dominate a mailed catfish community. Differentiating the sexes is difficult in long-snouts and best performed on the basis of the form of the ventral fins.

*Corydoras simulatus*, ♂

*Corydoras simulatus*, ♀

# Born mimics

*Corydoras sp. "Sychri-Longnose"*

*Corydoras leopardus*

*Corydoras cervinus*

*Corydoras maculifer*

*Corydoras sp. "Reynoldsi-Longnose"*

*Corydoras ellisae*

## Poisonous spines

Mailed catfishes really are harmless little chaps. But the lovely *Corydoras sterbai* is not altogether popular with those who work in the aquarium trade worldwide. A prick from its brilliant orange pectoral fin spines is not something you forget in a hurry! This appears to be a common feature of the species with eye-catching fin spines. A prick from *C. gossei* is also unpleasant, and I am told the same is true of *C. haraldschultzi*. We have already discussed self-poisoning in *C. pulcher*, but I haven't yet come across anyone who has been pricked by it, nor do I know anybody who has been pricked by *C. robustus*, which can also develop prolonged fin membranes in the dorsal fin.

It appears that in each case these colorful fin spines are warning signals. Perhaps these mailed catfish species occur in areas where predatory fishes like to grab a cory as a night-time snack. Perhaps this has led to a reinforcement of their defenses. Perhaps this is why they have become poisonous.

The breeding of a number of species in this group (*C. sterbai*, *C. gossei*, *C. haraldschultzi*) takes place regularly, and, because these fishes are very attractive and command high prices in the aquarium trade, they are also bred by professional breeders. For this reason these lovely fishes are available all the time. These three species seem to me to be a sister group to the other two (*C. robustus* and *C. pulcher*). Their maintenance is not difficult, except that people who suffer allergic reactions should avoid getting pricked. *C. robustus* grows rather large (about 9 cm), and, at about 6 cm, the other species are not the smallest of corys.

*Corydoras sterbai*

*Corydoras haraldschulzi*

*Corydoras robustus*

*Corydoras gossei*

*Corydoras seussi*

# All good things......

......must come to an end, as the saying goes. And so, regrettably, must our expedition through the world of mailed catfishes. I would have loved to talk about the sand-burrowing members genus *Aspidoras*. Or the splendid emerald catfishes of the genus *Brochis*, which can grow up to 15 cm long! Or the species pair *Corydoras caudimaculatus* and *C. similis* with their caudal peduncle spots. There is no space for the *Callichthynae*, the bubblenest-building armored sausages, nor for the dwarf forms among them. Even more commonly available species, such as *Corydoras robinae*, the species with the flag-tail, can only be mentioned in passing. Space is always limited in a book of this type. But perhaps the groups that have been covered will have served to show you that these fishes are in no way just useful, but boring, scavengers for the aquarium.

It is perhaps fitting that the final group portrayed here should be that with which this book began: the peppered catfish, *Corydoras paleatus*. Close relatives of this species include *C. erhardti*, *C. flaveolus*, *C. garbei*, and others. Anyone with the chance to buy wild-caught specimens of these species should snap them up, as such opportunities are rather rare. The majority of these fishes come from relatively southerly regions and are optimally maintained at room temperature, i.e. between 18 and 22 °C.

There is a whole series of cultivated forms of our old friend *C. paleatus*. Males of this species group are naturally significantly slimmer than females and have greatly enlarged fins. Particularly large-finned variants have been produced through selective breeding, often sold under the label "*C. steindachneri*" (which is another species entirely, probably never yet imported). In addition there are albinos (as an aside, there is also an albino form of *C. aeneus*), a gold form, and one with veil-fins. As to whether these captive-bred forms are necessary or not, more attractive or less than the natural form – in the final analysis it is all a matter of personal taste. It remains to say, in conclusion, never cross fishes from different localities with one another. In this world of ours, where extensive environmental destruction is the order of the day, it may happen – and rather more quickly than we anticipate – that we may need to give back to Nature what she has so generously given us for our aquaria. But only pure-blooded stocks will do!

*Corydoras paleatus*

*Corydoras erhardti*

# Breeding and rearing

The breeding of mailed catfishes has often been mentioned in earlier chapters. Some species can only with difficulty be prevented from breeding in the aquarium. These include the bronze catfish (*Corydoras aeneus*), Sterba's catfish (*C. sterbai*), and the peppered catfish (*C. paleatus*). In the case of these fishes a partial water change (about 40% of aquarium volume) is often enough to trigger spawning. Other species are appreciably more difficult to stimulate. In nature it is often rainy periods or seasonal changes in the food supply that induce ripening of the gonads. A very large number of species can be stimulated by performing no water changes at all for several weeks as soon the females' spawning tubes are apparent. At the end of this phase several large (about 80% of tank volume) water changes are carried out in quick succession, using cooler water (about 18 °C). Often spawning then follows.

Every species has its peculiarities. Some species lay a few eggs regularly over a long period of time, others a larger clutch all at one go. There are group-spawners and species where single pairs separate from the group. This must be taken into consideration in each case. For species that live in heavily shaded forest streams in the wild, the aquarium must be made as dark as possible for breeding, leaving just a viewing slit at the front so the activities of the fishes can be observed.

Often the fertility rate is poor. In such cases it may help to use soft, slightly acid water for the triggering water changes. The sperm of many fish species is barely capable of movement in hard alkaline water – this has not been proven for mailed catfishes, but seems likely. The further from the Equator a species originates, the more important is seasonal variation in temperature and water levels, and this too must be taken into account for breeding.

All mailed catfishes adopt a so-called T-position during spawning. This involves the male clasping the barbels of the female with his pectoral spines, while the female expels one or more eggs into a pouch formed by her folded ventral fins where they are fertilized in some manner that has yet to be explained. The inner side of the pectoral fin spines is characteristically toothed in each species, and this probably serves to prevent sympatric species from crossing. Unfortunately this mechanism doesn't seem to work in the aquarium, and so similar species should not be kept together.

The use of spawning mops in a special breeding tank is a proven technique. In the absence of any other spawning substrate the females will often stick their eggs to these, and they can then be easily collected by the breeder. In many species, especially the long-snouts, a strong current must be provided in the aquarium. Some species eat their spawn, others ignore it. Often, however, it is non-spawning conspecifics that attack the eggs.

The majority of breeders remove the spawn from the aquarium and place it in small hatching containers where the water must always be kept clean, i.e. changed daily. Dead (= white) eggs must be removed. Hatching takes place after 3-5 days, and the yolk sac is absorbed within another 2-3 days. Fry of all species can manage *Artemia* nauplii right from the start. Most losses among fry result from bacterial attack and/or proliferation of infusorians in too crowded containers. I personally advise rearing the fry for the first few weeks in small tanks with a sandy bottom and low water level (about 5 cm). The water must be changed daily and dirt removed using an airline siphon.

# Eclipse™ Aquarium System

## ADVANCED TECHNOLOGY, SUPERIOR PERFORMANCE, UNLIMITED VERSATILITY...

**Superior BIO-Wheel Filtration**
Silent, high capacity 3-stage efficiency. BIO-Wheel and Eclipse Filter Cartridge unmatched by all other types of aquarium filtration.

**Superior Illumination**
Colour-enhancing fluorescent lighting. Far better that heat-producing incandescent bulbs. Plants thrive and colours of fish and plants come alive!

**Superior Convenience**
Easy set-up, easy operation. Polymer wool and carbon all-in-one filter cartridge changes in seconds, whilst the BIO-Wheel never needs replacing.

**Superior View**
Injection-molded acrylic aquarium provides a panoramic 360 degrees of prime viewing area for maximum enjoyment. Available as *Eclipse System 3* and *Eclipse Systems 6*.

*Eclipse Explorer*
Employs the same sophisticated BIO-Wheel filtration technology as the larger Eclipse Systems. Available in 4 additional fun colours, incorporating a textured skylight to maximise surrounding light.

Explorer = 7.5 litres    System 3 = 11 litres    Sytem6 = 22.5 litres

ALL AQUALOG TITLES & THE ABOVE PRODUCTS ARE DISTRIBUTED IN THE UK BY:

Belton Road West ~ Loughborough
Leicestershire ~ LE11 5TR
Tel: 01509 610310 ~ Fax: 01509 610304
E-mail: info@underworldproducts.co.uk
Web Site: www.underworldproducts.co.uk

# Textbooks – detailed guides to maintenance and breeding

**Most advisories include a beautiful poster!**

- Detailed guidance on maintenance and breeding, tricks and tips from experienced specialists
- Many volumes include a decorative color poster (85 x 60 cm, also available separately)
- Available in German and English editions

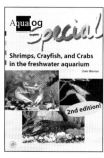

**Shrimps, Crayfish and Crabs in the Freshwater Aquarium**
(U. Werner)

We present the most beautiful freshwater crustaceans. Small and large species that can be kept alone or with fishes – but which? The answer, and much more, in this interesting and engrossing guide.

(64 pages)
ISBN 3-936027-08-0
Item no. AS010-E

**Decorative Aquaria The Beginner´s tank**
(U. Glaser sen.)

For the first time a detailed guide to setting up a perfect, beautiful, aquarium for the novice. Expert advice on avoiding beginners' mistakes. Tank layout, plant and fish populations; all illustrated, and described in detailed but easy-to-understand terms.

(48 pages + poster)
ISBN 3-931702-38-3
Item no. AS011-E

**The Natural Garden Pond**
(P. D. Sicka)

A garden pond that imitates nature is a refuge for innumerable endangered animals and plants. Numerous examples and splendid photos demonstrate clearly how to create a dream of a miniature biotope in your own garden.

Available in German only.

(48 pages)
ISBN 3-931702-90-1
Artikel-Nr. AS021-D

**The most beautiful L-numbers**
(U. Glaser sen.)

What are L-catfishes and where do they come from? Expert advice on maintenance and breeding, etc.

(48 pages + poster)
ISBN 3-931702-33-2
Item no. AS002-E

**Fascinating Koi**
(H. Bachmann)

A summary of the long history of colored carp, accurate guidance on maintenance and what a koi pond should be like. Plus much more useful information from an expert.

(48 pages + poster)
ISBN 3-931702-41-3
Artikel-Nr. AS003-E

**Freshwater Coral Fish MALAWI cichlids**
(E. Schraml)

Colorful as marine fishes, but requiring far less outlay on equipment for maintenance and breeding. Learn the secrets of success in this guide by an expert.

(48 pages + poster)
ISBN 3-931702-49-9
Item no. AS009-E

**Goldfish and Fancy Goldfish**
(K. H. Bernhardt)

The oldest and best-known ornamental fish, but do you realize how many forms and color varieties exist? Interesting facts on the history of these fish, plus numerous tips on their care, as they are not as hardy as often assumed.

(48 pages + poster)
ISBN 3-931702-45-6
Item no. AS008-E

**Fishes of the Year The HIGHLIGHTS**
(U. Glaser sen.)

Every year new fishes appear in the trade – that is what makes the aquarium hobby so exciting. All about their origin, wild or tank-bred, awards, and characteristics of these fishes. Quick, up-to-date information for every aquarist.

(48 pages + poster)
ISBN 3-931702-69-3
Item no. AS007-E

**Breathtaking Rainbows**
(H. Hieronimus)

The name says it all: all the colors of the rainbow. A guide to maintenance and other essential knowledge. The biotope photos show where these pretty, easy-to-keep fishes come from and how to set up a suitable aquarium.

(48 pages + poster)
ISBN 3-931702-51-0
Item no. AS004-E

**Majestic Discus**
(M. Göbel)

The king of fishes, dream of every aquarist! The care of these demanding fishes and lots more advice from an expert.

(48 pages + poster)
ISBN 3-931702-43-X
Item no. AS006-E

**Freshwater Stingrays from South America**
(R. A. Ross)

The rivers and lakes of South America are home to fishes more usually associated with the sea: stingrays. Although feared in their home waters on account of their poisonous spines, these unique fishes have caught the imagination of increasing numbers of aquarists worldwide. This book is the first comprehensive guide to the successful maintenance and breeding of these rays. Indispensable for anyone wishing to learn more about these interesting creatures or attempt to keep them.

(64 pages)
ISBN 3-931702-87-1
Item no. AS013-E

# Pictorial lexicons –
## all known fishes in a group

**All the important information at a glance:**

- Color photos of all the fishes of a group (inc. all varieties, color and cultivated forms)
- Identification of any fish is accurate and easy: scientific name, hobby name, Aqualog code number
- Easy-to-understand text, international maintenance symbols
- Newly discovered fishes published as supplements: your lexicon will always remain up-to-date!

*These 3 pictorial lexicons form a compact identification guide, for the first time covering all the killifishes of the world: the official reference book for killifish fans worldwide!*

**Killifishes of the world – Old World Killis I**
(L. Seegers)

Killies are also known as freshwater jewels – if you have seen their splendid colors you will know why. This volume covers the genus Aphyosemion plus the lamp-eyes and ricefishes.

(160 pages, more than 890 color photos)
ISBN 3-931702-25-1
Item no. B007

**Killifishes of the world – Old World Killis II**
(L. Seegers)

Volume 2 covers the Nothobranchius, Epiplatys, Aplocheilus, Aphanius, and others. Their colors and size (only 3-8 cm) make them ideal aquarium fishes.

(112 pages, 550 color photos)
ISBN 3-931702-30-8
Item no. B008

**Killifishes of the world – New World Killis**
(L. Seegers)

This book completes the killifish series with groups from the New World: Rivulus, Cynolebias, Fundulus, Pterolebias, and others.

(224 pages, 1200 color photos)
ISBN 3-931702-76-6
Item no. B014

**The Puffers of fresh and brackish waters**
(K. Ebert)

Not only 300 brilliant photos of all the puffers of the world, but more than 40 years of experience maintaining these unusual colorful creatures are shared by the author in this unique lexicon intended for the novice as well as specialist aquarists and scientists.

(96 pages, 300 color photos)
ISBN 3-931702-60-X
Item no. B016-E

*All the cichlids of Latin America in 4 volumes!*

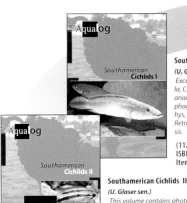

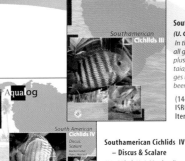

**Southamerican Cichlids I**
(U. Glaser sen.)
Excellent color photos of all Cichla, Crenicichla, Teleocichla, Guianacara, Geophagus, Gymnogeophagus, Satanoperca, Acarichthys, Uaru, Biotodoma, Astronotus, Retroculus, and Chaetobranchopsis.

(112 pages, 500 color photos)
ISBN 3-931702-04-9
Item no. B002

**Southamerican Cichlids II**
(U. Glaser sen.)
This volume contains photos, by well-known photographers, of all dwarf cichlids including Apistogramma, Biotoecus, Crenicara, Dicrossus, Nannacara, Taeniacara, and Microgeophagus (formerly Papiliochromis).

(112 pages, 500 color photos)
ISBN 3-931702-07-3
Item no. B003

**Southamerican Cichlids III**
(U. Glaser sen.)
In this volume you will find the catchall genera Aequidens and Cichlasoma, plus the allied genera Acaronia, Caquetaia, Petenia, and Herotilapia. Changes in scientific names since 1988 have been taken into account.

(144 pages, 650 color photos)
ISBN 3-931702-10-3
Item no. B005

**Southamerican Cichlids IV – Discus & Scalare**
(M. Göbel, H. J. Mayland)
Volume 4 covers the fantastic discus and angelfishes. Wild-caught plus German, other European, and Asian captive-bred fishes, including all varieties, color sports, and cultivated forms.

(240 pages, more than 900 color photos)
ISBN 3-931702-75-8
Item no. B010

**Freshwater Rays**
(R. A. Ross / F. Schäfer)

All known species of freshwater rays in all their vast variety. For the first time in the history of aquarium literature a reference work including the South American fluviatile rays (Potamotrygonidae) plus the Asian, African, North American, and Australian freshwater species. Also includes the sawfishes (Pristidae) and ray species regularly found in brackish water all over the world.

(192 pages, 400 approx. color photos)
ISBN 3-931702-93-6
Item no. B015

# Pictorial lexicons –
## all known fishes in a group

**All the important information at a glance:**

- Color photos of all the fishes of a group (inc. all varieties, color and cultivated forms)
- Identification of any fish is accurate and easy: scientific name, hobby name, Aqualog code number
- Easy-to-understand text, international maintenance symbols
- Newly discovered fishes published as supplements: your lexicon will always remain up-to-date!

**All rainbows**
(H. Hieronimus)

All the colors of the rainbow, as the name implies. All species known to date are to be found in this book, but many more remain to be discovered in biotopes, eg in Papua New Guinea, where collecting is extremely difficult.

(144 pages, 700 approx. color photos)
ISBN 3-931702-80-4
Item no. B013

**All livebearers**
(M. Kempkes, F. Schäfer)

For the first time all the livebearers are illustrated – the well-known guppy, mollies, swordtails, platies, plus all the others. All the wild and cultivated forms/color varieties, as well as the halfbeaks.

(352 pages, 2000 approx. color photos)
ISBN 3-931702-77-4
Item no. B009

**All Corydoras**
(U. Glaser sen.)

All the mailed catfishes are presented together for the first time. As well as the genera Aspidoras, Brochis, Callichthys, Corydoras, Dianema, and Hoplosternum, there are also all variants, mutants, hybrids, cultivated forms, and undescribed species ("C-No").

(144 pages, 650 color photos)
ISBN 3-931702-13-8
Item no. B004

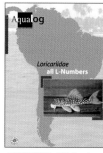

**LORICARIIDAE – All L-Numbers**
(U. Glaser sen.)

All L-number catfishes up to L204, with the rest as supplements. The only book to illustrate and describe all the L-number catfishes.

(112 pages, 450 approx. color photos)
ISBN 3-931702-01-4
Item no. B001

**African Cichlids I MALAWI MBUNA**
(E. Schraml)

This book really does show all mbuna species and variants covered in the lake to date!

(240 pages, 1500 approx. color photos)
ISBN 3-931702-79-0
Item no. B012

**All Goldfish and Varieties**
(K. H. Bernhardt)

The goldfish is the oldest ornamental fish in the world, familiar to everyone - but how many people know that there are so incredibly many varieties? This pictorial lexicon includes all the forms and color variants.

(160 pages, 690 color photos)
ISBN 3-931702-78-2
Item no. B011

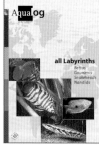

**All Labyrinths**
(F. Schäfer)

For the first time a compact lexicon illustrating all the labyrinth fishes. Plus the snakeheads, nandids, Pristolepidae and Badidae which exhibit many behavioral parallels with the labyrinths. Also includes an identification key to the genus Betta. The official reference guide for labyrinthfish societies worldwide as soon as it was published.

(144 pages, 690 color photos)
ISBN 3-931702-21-9
Item no. B006

## Aqualog Photo Collection

### Book + CD-ROM

**All books incl. CD-ROM**

- This series of books portrays fishes of various groups using top-quality color photos
- Unambiguous identification via international code numbers, scientific and hobby names
- Brief details: Characteristics, maintenance requirements, etc.
- All photos in each book also on the accompanying CD-ROM

Version A: German, Japanese, Czech, Turkish, Hungarian
Version B: English, Dutch, Swedish, Danish, Finnish
Version C: French, Spanish, Italian, Polish, Mandarin

**Photo collection No. 1**
(U. Glaser sen.)
**African catfishes**
A: ISBN 3-931702-56-1
B: ISBN 3-931702-57-X
C: ISBN 3-931702-58-8
Item no. PC001-A/B/C

**Photo collection No. 2**
(U. Glaser sen.)
**Characins 1**
(African characins, predatory characins, pencilfishes)
A: ISBN 3-931702-59-6
B: ISBN 3-931702-62-6
C: ISBN 3-931702-63-4
Item no. PC002-A/B/C

**Photo collection No. 3**
(U. Glaser sen.)
**Characins 2**
(Piranhas, silver dollars, headstanders, hatchetfishes)
A: ISBN 3-931702-64-2
B: ISBN 3-931702-65-0
C: ISBN 3-931702-66-9
Item no. PC003-A/B/C

**Photo collection No. 4**
(U. Glaser sen.)
**Characins 3**
(Neons, Moenkhausia, American predatory characins, characins)
A: ISBN 3-931702-81-2
B: ISBN 3-931702-44-8
C: ISBN 3-931702-47-2
Item no. PC004-A/B/C

**Each volume contains 96-112 pages and approx. 300-400 color photos**